SpringerBriefs in Physics

SpringerBriefs in Physics are a series of slim high-quality publications encompassing the entire spectrum of physics. Manuscripts for SpringerBriefs in Physics will be evaluated by Springer and by members of the Editorial Board. Proposals and other communication should be sent to your Publishing Editors at Springer.

Featuring compact volumes of 50 to 125 pages (approximately 20,000–45,000 words), Briefs are shorter than a conventional book but longer than a journal article. Thus, Briefs serve as timely, concise tools for students, researchers, and professionals.

Typical texts for publication might include:

- A snapshot review of the current state of a hot or emerging field
- A concise introduction to core concepts that students must understand in order to make independent contributions
- An extended research report giving more details and discussion than is possible in a conventional journal article
- A manual describing underlying principles and best practices for an experimental technique
- An essay exploring new ideas within physics, related philosophical issues, or broader topics such as science and society

Briefs allow authors to present their ideas and readers to absorb them with minimal time investment. Briefs will be published as part of Springer's eBook collection, with millions of users worldwide. In addition, they will be available, just like other books, for individual print and electronic purchase. Briefs are characterized by fast, global electronic dissemination, straightforward publishing agreements, easy-to-use manuscript preparation and formatting guidelines, and expedited production schedules. We aim for publication 8–12 weeks after acceptance.

SpringerBriefs in Physics

SpringerBriefs in Physics are a series of slim high-quality publications encompassing the entire spectrum of physics. Manuscripts for SpringerBriefs in Physics will be evaluated by Springer and by members of the Editorial Board. Proposals and other communication should be sent to your Publishing Editors at Springer.

Featuring compact volumes of 50 to 125 pages (approximately 20,000–45,000 words), Briefs are shorter than a conventional book but longer than a journal article. Thus, Briefs serve as timely, concise tools for students, researchers, and professionals.

Typical texts for publication might include:

- A snapshot review of the current state of a hot or emerging field
- A concise introduction to core concepts that students must understand in order to make independent contributions
- An extended research report giving more details and discussion than is possible in a conventional journal article
- A manual describing underlying principles and best practices for an experimental technique
- An essay exploring new ideas within physics, related philosophical issues, or broader topics such as science and society

Briefs allow authors to present their ideas and readers to absorb them with minimal time investment. Briefs will be published as part of Springer's eBook collection, with millions of users worldwide. In addition, they will be available, just like other books, for individual print and electronic purchase. Briefs are characterized by fast, global electronic dissemination, straightforward publishing agreements, easy-to-use manuscript preparation and formatting guidelines, and expedited production schedules. We aim for publication 8–12 weeks after acceptance.

George Musser

Emergence in Condensed Matter and Quantum Gravity

A Nontechnical Review

 Springer

George Musser
Glen Ridge, NJ, USA

ISSN 2191-5423   ISSN 2191-5431   (electronic)
SpringerBriefs in Physics
ISBN 978-3-031-09894-9   ISBN 978-3-031-09895-6   (eBook)
https://doi.org/10.1007/978-3-031-09895-6

This Springer imprint is published by the registered company Springer Nature Switzerland AG
The registered company address is: Gewerbestrasse 11, 6330 Cham, Switzerland

For Ann and Ben

Acknowledgements

This review was originally prepared by the author working with the Foundational Questions Institute, FQXi, for the John Templeton Foundation. Anthony Aguirre, Adam Brown, David Tong, and Mark Van Raamsdonk generously took the time to read portions of the draft and provide detailed feedback. I started to collect material for this review while a Director's Visitor at the Institute for Advanced Study and would like to thank Robbert Dijkgraaf, Nima Arkani-Hamed, and Josephine Faass for making that possible. Over the years, I have been privileged to learn from the world's leading physicists and philosophers of physics, who are a remarkably patient and good-humored lot. Unable to list them all here, I thank them as a community. Elfy Chiang created the Quantum Hall Effect graphic in Chap. 2, and Maayan Harel developed the spacetime-holography graphic in Chap. 3—how lucky I was to work with such talented artists. I am delighted to have come to know Thomas Burnett from the John Templeton Foundation, who coordinated the FQXi/JTF reviews. I have been honored to work with Angela Lahee, Ashok Arumairaj, and Sindhu Sundararajan at Springer to translate the Templeton project into a hardcopy volume. And Zeeya Merali—I have known her for much of my career and never ceased to learn from her incisive writing and editing. I can think of no more knowledgeable editor I could have worked with on this project.

Contents

About the Author

George Musser is an award-winning journalist, a contributing editor for *Scientific American*, a contributing writer to *Quanta*, and the author of two books on fundamental physics, *The Complete Idiot's Guide to String Theory* and *Spooky Action at a Distance*. He is the recipient of an American Astronomical Society's Jonathan Eberhart Planetary Sciences Journalism Award and the 2011 American Institute of Physics Science Writing Award. He was a Knight Science Journalism Fellow at MIT from 2014 to 2015. He has appeared on *The Today Show*, CNN, NPR, the BBC, and more. He lives in Glen Ridge, NJ, with his wife, daughter, and Schnauzer. Follow him on Twitter at @gmusser.

1

Introduction

Abstract The universe is not just a formless blob of particles. It has structures on multiple scales. Why? You might expect vast numbers of particles to behave like vast numbers of particles, far too complex for any mortal mind to grasp. Yet they achieve a collective simplicity. Scientists say higher-level properties "emerge." Exactly how they do this is still one of the great questions of science. It is not even clear whether emergence merely repackages microscopic physics or adds something genuinely new. Every field of science contributes to the study of emergence. The same principles of organization that govern masses of particles also apply to bird flocks, traffic jams, and national economies. But if any discipline can claim emergence as its central concern, it is condensed-matter physics, which analyzes how material properties flow from fundamental physics. Its practitioners have discovered entirely new forms of collective order involving quantum physics and the special mathematics of topology. Their insights are migrating even into fundamental physics. Theorists studying quantum gravity now think that space and time, once thought the rock-bottom level of reality, are emergent structures from an even deeper one.

Keywords Emergence · Emergent properties · Reductionism · Weak emergence · Strong emergence

© The Author(s), under exclusive license to Springer Nature
Switzerland AG 2022
G. Musser, *Emergence in Condensed Matter and Quantum Gravity*,
SpringerBriefs in Physics, https://doi.org/10.1007/978-3-031-09895-6_1

1.1   What is Emergence?

Some things are clearly complex, but it's the things that look simple that you have to watch out for. We know better nowadays than to think of the air in a room as empty space, but still it seems bland. Dust motes and aerosol particles float through it, and scents are carried on the breeze, but that is about the most complexity we ascribe to air. Even the most imaginative physicist can scarcely envision the countless little dramas playing out as molecules bounce and ram and swirl. There is more interaction in a thimbleful of air than across the entire Internet, and more in one second than in all of human history.

It comes as no surprise that trillions of atoms are complicated. What is much stranger is that they should ever look simple. Stranger still, the higher-level simplicity can be studied on its own. To build the steam engines that powered the industrial revolution, engineers did not have to know about molecules. It was enough to traffic in quantities such as pressure, density, and temperature. Only later did physicists recognize these quantities as the collective properties of molecules confined in a chamber. For instance, when air is warm, molecules move faster on average than they do when air is cold. Yet it makes no sense to speak of the temperature of an individual molecule or to identify temperature with the speed of just that molecule. Even in warm air, some molecules move quite slowly; it is only the overall distribution of speeds—a collective property—that determines the temperature. Similarly, pressure and density follow their own logic that does not require you to know about the molecules. There is, thankfully, no need to track all those molecules individually to know what they will do in bulk.

The outward simplicity is not in contradiction with the inner complexity; to the contrary, one emerges from the other. How complexity begets simplicity is still one of the great questions of science. It often seems, even to the experts, like a magic trick: here are a bunch of atoms and, hey presto, they do collectively what none could individually. The need to articulate principles of organization is obvious in biology, since concepts such as Darwinian evolution and self-assembly are nowhere to be found in fundamental physics. But it also applies even within physics. The systems that physicists study typically involve processes occurring on multiple scales, and a single mathematical model doesn't capture that (Green & Batterman, 2017). In shock waves, for example, wave propagation and viscous dissipation are both important, yet differ vastly in their scale (Saatsi & Reutlinger, 2018). That means they cannot be reduced to the microlevel dynamics. The shock requires a higher-level view.

Gradually, physicists are pulling back the curtain on how the natural world orders itself. The past decade has seen a flowering of work on emergence in numerous disciplines, and this review will cover advances in two major fields of study: condensed-matter physics—the study of the structure and behavior of matter—and the search to understand the origins of spacetime. Chapter 2 discusses the paradigmatic example of emergence within physics: changes of state. Even familiar changes of state, such as liquids freezing into solids or boiling into gases, involve highly elaborate choreographies of vast numbers of molecules. Physicists now refer to solids, liquids, and gases as the three classical phases of matter. But these phases do not begin to exhaust the capabilities even of seemingly ordinary systems.

Condensed-matter physicists have now discovered a wealth of new, distinct, and sometimes mysterious additional phases. Systems enter one of these phases when their temperature changes or other external conditions are varied. In recent years, physicists and material scientists studying these materials have catalogued a number of bizarre properties that have never been observed before. Such properties are described as emergent because they could not have been predicted by looking at the arrangement of the matter's constituent atoms alone. Physicists have been also homing in on the mechanisms behind some of these exotic emergent properties. This is leading to new practical applications and a better understanding of foundational physics.

The second chapter will cover three broad categories of phases that are generating huge interest among condensed-matter physicists: soft matter, quantum phases of matter, and topological phases. Soft matter comprises some familiar materials, from plastic to sand grains, liquid crystals to body tissue, that have only recently been found to share some strange characteristics. The macroscopic properties of soft matter cannot be understood by looking at the microscopic arrangement of their particles; rather you must look at order at an intermediate scale. Their complex behavior arises as a result of self-assembly. Understanding these mechanisms is important for developing new organic electronics for computer displays and smartphones, as well as for cancer therapy and organ regeneration in medicine, among other applications.

Quantum phases of matter encompass a vast range of materials from the superconductors that are now routinely used in hospital scanners to less widely known materials produced in research labs, such as Bose–Einstein condensates and strange metals, which will no doubt find their own technological applications. Although superconductors were discovered over a century ago, physicists are still trying to solve the mystery of how different types of superconducting behaviors occur. To understand how the electrons in certain materials can flow without resistance, under specific conditions, they have turned to quantum physics. Developed a century ago to describe the very small—particles and atoms—quantum theory turns out to scale up to large systems as well, these new materials being the latest examples. Quantum physics describes how, at extremely low temperatures, particles will wiggle wildly as you try to confine them, eventually leading them to re-organize, creating new and useful emergent phenomena.

Another hugely important quantum feature is entanglement—a special link that can connect and coordinate the motion and behavior of particles across materials. Physicists and material scientists are beginning to explore how these quantum properties can be exploited to create desirable electronic features in the novel wonder-material graphene—a particularly strong form of carbon that can be easily manipulated to create better quality transistors, sensors, and electronic components.

The final category of materials discussed in Chap. 2 involves topological phases, which have risen in prominence over the last decade or so. Like quantum phases they can feature long-range ordering of particles across a material. In addition, particle motions can become coordinated in extremely complex ways. As a result, topological materials play host to bizarre and attractive electronic properties. For example, topological insulators conduct electricity on their edges, but serve as electrical insulators in their interior. Such materials could potentially be used to build revolutionary new electronic devices or construct robust error-resistant quantum computers—machines that should in theory dramatically outperform everyday classical computers at certain specific tasks.

Topological materials are so new that physicists are discovering more candidate materials each day; the chapter also describes the role that artificial intelligence (A.I.) and machine learning can play in this hunt. Finally, Chap. 2 describes an ambitious new unifying framework that attempts to describe space itself as a topological phase called a string-net liquid. Such a model could explain how elementary particles, and even light and the electromagnetic force, emerge.

This broad search for a theory of everything—or, specifically, a framework that combines quantum theory, originally developed to describe atoms and subatomic particles, with gravity—is examined in more depth in Chap. 3. Emergence has become a central theme in this endeavor. At first, it seemed like nothing could be much simpler than space. Almost by definition, it is sheer emptiness, merely a container for matter and energy. But, as described in Chap. 3, Einstein realized that space is woven together with time to create a spacetime fabric that pervades the universe, and this is a dynamic entity. Since then, physicists have calculated that space has an enormous latent complexity, more than anything else known to science. It even makes air, with its trillions upon trillion of molecules, seem pitifully austere.

Over the past few decades there have been mounting hints that space itself emerges from quantum processes. The first comes from a detailed theoretical analysis of the nature of black holes and the fate of objects swallowed by these cosmic monsters. The paradoxes black holes pose suggest that they are made from as yet unknown building blocks, which could serve as "atoms" of space in general. A second clue comes from quantum entanglement, which suggests that the odd linkage seen in labs between particles cannot be explained by processes operating within space and time. Together, these ideas have led some physicists to posit that space emerges from entanglement among the atoms of space.

In particular, the third chapter focuses on the holographic principle, which suggests that our three-dimensional space may emerge from quantum inter-actions on a two-dimensional surface. In some versions of the principle, quantum interactions do not occur in any space at all, bootstrapping the universe from scratch. This concept is closely tied to a conjecture from the 1990s (called the "Anti de Sitter/Conformal Field Theory," or AdS/CFT, duality) which says that the mathematics used to calculate the behavior of black holes and other exotic objects in a hypothetical kind of 3-D space is equivalent to that used to describe quantum interactions between particles in a 2-D space. Since then, various physicists have devised mechanisms whereby space can be stitched together by quantum entanglement and wormholes— tunnels in space and time—and how gravity may also emerge from such a picture. Others have re-conceived the AdS/CFT relationship in analogy with a computer error-correcting code to both sharpen how space emerges and, perhaps, resolve the black-hole information paradox. Such ideas may also be tested, in some capacity, in table-top lab experiments.

The AdS/CFT duality is popular, but is by no means the only attempt to explain the origins of space. Chapter 3 also reviews some well-regarded alternative models, including loop quantum gravity, in which gravity emerges from a set of geometric relations; casual set theory, in which spacetime is constructed from a network of cause-and-effect relationships; and causal dynamical triangulations, in which spacetime assembles itself from a mélange of all possible shapes.

Finally, Chap. 4 describes how theoretical developments in understanding the emergence of space have fed back into condensed matter, enabling scientists to pool their understanding and solve the problems of one area by importing ideas from the other. The earliest examples that resonances between condensed matter and quantum gravity could have real practical results came from physicists applying the holographic principle to explain and predict the behavior of nuclear plasmas produced in particle accelerators. In the early part of the twenty-first century, they found inspiring results that AdS/CFT could be used to describe plasma viscosity. Others then successfully applied the framework to superconductivity and other effects seen in quantum phases.

This interdisciplinary research program has led to the development of the so-called SYK model, which was first applied within condensed-matter physics to strange metals and was then turned towards analyzing the emergence of space. The model has contributed to efforts to resolve the black-hole information paradox, and it offers the possibility of testing abstract theories of the emergence of space in the lab.

While physicists are at the heart of research into emergence, philosophers of science have been drawn in, too—not just because emergence is important, but because it can be confusing. The word "emergence" is used in multiple ways, often causing scholars to talk past one another. Philosophers have sought clarity. The remainder of this introductory chapter covers some of the conceptual debates about what emergence really means.

1.2   Is Emergence Fundamental?

For decades, emergence was considered the opposite of reduction. Physicists are often said to break down or "reduce" the world to its parts, and physics can be defined as the search for simple laws that underlie the seeming anarchy of nature. Their methods contrast with the more synthetic style of sciences such as biology. Physicists became the butt of jokes about "spherical cows"—idealizing the world until it ceased to look like the real world—while physicists disparaged other disciplines as "stamp-collecting"—merely categorizing without providing any deeper explanation.

These scholarly fault lines were never very sharp and the dichotomy now seems passé. Emergence and reductionism are two sides of a coin. Emergence is meaningless unless there are parts to assemble, and reductionism is meaningless unless you can explain what those parts do. Any good set of IKEA instructions lists both the parts and the assembly steps. Even particle physicists, whose entire field seems to be about reduction, spend at least as much time on the constructive side. That aspect to their work is less known perhaps because it is harder to visualize. Stuff is made of atoms and particles—that's easy to grasp. But who outside physics has heard of the renormalization group? It and other methods to understand complexity don't lend themselves to easy encapsulation. But they are no less compelling for it.

That said, a trace of this old dispute lives on in other questions. One that fascinates philosophers of science is whether emergence is an objective feature of the world or a product of our own limitations as observers and theorizers. As Bedau (2008) has put it, is "emergence just in the mind"?

Those who argue that emergence is just a human construct note that the fundamental laws of physics are all-encompassing; they describe the world in full generality. We invoke emergent laws, such as those of biology, only when we confine attention to a slice of the world or a particular sequence of events. Seager (2012, p. 183) argues that higher levels are relational: they express the relations among the basic building blocks, but, like any relation, do not exist per se: "They are rather like the clouds that appear in the shape of animals." Taylor (2014, 2015) argues that levels are "gerrymandered"—arbitrary distinctions. To the extent they are real, they reflect the types of explanations we have available to us.

Put simply, these philosophers argue that emergence is about us. We speak of it only when, for practical reasons or simply for convenience, we neglect microscopic details. But Seager also acknowledges a clear problem with this argument: us. Our own minds are emergent from unconscious atoms. If emergence is about us, then it follows that there isn't really any "us," and we go round and round in a logical loop (Seager, 2012).

Others, such as Bedau and Batterman (2017), and Green (2017), have broader arguments about the objective nature of emergence. They argue that you really do need to move to the collective level—that some aspects of the behavior of a system are not present at the base level, even in a highly convoluted form. That, in turn, poses the question of whether the emergence should be described as "weak" or "strong."

1.3 Weak Versus Strong Emergence

The structure of the world has a tension that is familiar to parents of teenagers: dependence versus autonomy. When considering an emergent property, you can ask, is this feature derived from and dependent on the interactions of its constituent parts? Or is the whole greater than the sum of its parts, with features that are an independent element of reality?

The weak variety of emergence is fairly uncontroversial. It gives as much weight to dependence as autonomy. In principle, you could derive weakly emergent properties from the base layer. The examples of pressure, density, and temperature are weakly emergent; we may not need to know the details of all the molecules that give rise to these features, but the emergent properties are clearly a direct result of physics at the microscale.

Strong emergence, by contrast, tilts the balance further from dependence to autonomy. It asserts that genuinely new properties arise at the collective level that can't be derived from the fine details, period. While these higher-level features are compatible with fundamental physics, you could not predict they would arise by looking at the constituent parts, nor can studying the base level teach you about how those emergent properties will unfold.

A case study is chemistry. On the one hand, it seems like a clear example of weak emergence. The subject is firmly rooted in physics; chemical properties and reactions boil down to electron orbitals and intermolecular interactions. Some physicists, in their more intemperate moments, deem chemistry merely a branch of physics. On the other hand, Hendry (2017) has suggested that chemistry should be seen as strongly emergent. The equations for atoms and molecules beyond hydrogen are impossible to solve, which is not just a mathematical failure, but a signal that the atoms or molecules have genuinely novel properties. Chemical properties are defined in terms of their functions, not their microscopic composition, and indeed the microscopic details could vary yet produce the same effects (Manafu, 2015).

For some philosophers, strong emergence is tantamount to invoking the supernatural—a deviation from physics (Bedau, 2002). But others not only find it plausible, but see evidence of it in physics (McLeish, 2017). The next chapter will explore two possible sources of strong emergence: quantum entanglement and topology.

References

Batterman, R. W. (2017). Autonomy of Theories: An Explanatory Problem. *Noûs*, *52*(4), 858-873. https://doi.org/10.1111/nous.12191

Bedau, M. A. (2002). Downward causation and the autonomy of weak emergence. *Principia: an international journal of epistemology*, *6*(1). https://periodicos.ufsc.br/index.php/principia/issue/view/1392

Bedau, M. A. (2008). Is Weak Emergence Just in the Mind? *Minds and Machines*, *18*(4), 443-459. https://doi.org/10.1007/s11023-008-9122-6

Green, S., & Batterman, R. W. (2017). Biology meets physics: Reductionism and multi-scale modeling of morphogenesis. *Studies in History and Philosophy of Science Part C: Studies in History and Philosophy of Biological and Biomedical Sciences*, *61*, 20-34. https://doi.org/10.1016/j.shpsc.2016.12.003

Hendry, R. F. (2017). Prospects for Strong Emergence in Chemistry. In (1 ed., pp. 146–163). Routledge. https://doi.org/10.4324/9781315638577-9

Manafu, A. (2015). A Novel Approach to Emergence in Chemistry. In *Boston Studies in the Philosophy and History of Science: Philosophy of Chemistry* (pp. 39–55). Springer Netherlands. https://doi.org/10.1007/978-94-017-9364-3_4

McLeish, T. (2017). Strong emergence and downward causation in biological physics. *Philosophica*, *92*(2), 113-138.

Saatsi, J., & Reutlinger, A. (2018). Taking Reductionism to the Limit: How to Rebut the Antireductionist Argument from Infinite Limits. *Philosophy of Science*, *85*(3), 455-482. https://doi.org/10.1086/697735

Seager, W. (2012). *Natural Fabrications: Science, Emergence and Consciousness.* Springer. https://doi.org/10.1007/978-3-642-29599-7

Taylor, E. (2014). An explication of emergence. *Philosophical Studies*, *172*(3), 653-669. https://doi.org/10.1007/s11098-014-0324-x

Taylor, E. (2015). Collapsing Emergence. *The Philosophical Quarterly*, *65*(261), 732-753. https://doi.org/10.1093/pq/pqv039

2

Emergence in Condensed Matter Physics

Abstract Phase transitions such as melting and boiling are the paradigm for emergence in physics. They are the domain of condensed-matter physics. Researchers have discovered a bewildering variety of these transitions and have steadily enlarged their conception of what can happen during one. In broad terms, phase transitions occur when the balance of order and disorder shifts within a system. Quantum physics has given this basic insight a new slant, enabling remarkable phases of matter such as superconductors and strange metals. Topological phases such as the fractional quantum Hall effect go a step further and involve dynamics so rich that even experts have trouble visualizing it. Lately excitement has also been building around phases of matter that occur away from thermal equilibrium. Even household materials such as foam and rubber, classified under the new rubric of "soft matter," are governed by universal principles. Many of these phases are finding practical applications for sensors and computers, and the basic concepts are helping scientists in fields such as atmospheric science.

Keywords Condensed matter · Phase of matter · Phase transition · Mean-field theory · Symmetry · Soft matter · Superconductor · Superconductivity · Bose–Einstein condensate · Strange metal · Hall effectd · Topological insulator · String-net liquid · Many-body localization · Time crystal · Graphene · Neural network · Quantum entanglement

G. Musser, *Emergence in Condensed Matter and Quantum Gravity*,
SpringerBriefs in Physics, https://doi.org/10.1007/978-3-031-09895-6_2

2.1   Introduction

A humble tray of ice cubes puts human society to shame. It's hard to get even two people to agree on anything, yet a million billion water molecules can suddenly and abruptly coordinate to lock themselves into an ice crystal. On a winter's day, the steam from a whistling kettle condenses into droplets on a window—it seems humdrum, but imagine the countless molecules in the mist meeting and coagulating. Through their marvelous capacity for spontaneous organization, a few types of atoms make up almost everything we see. The bulk of life on Earth is made from six chemical elements. They do not merely combine into multitudinous chemical compounds, but assemble into a vast number of states of matter. And as physicists venture to realms of extreme cold or pressure, they find more and more states of matter that are alien to human experience. Solid, liquid, gas: that's just the start. This chapter will discuss some of physicists' more exotic and exciting discoveries.

When materials freeze or boil, they are said to undergo a phase transition. In most of the transitions we are familiar with, the density jumps and there is a period when both phases co-exist—water sits in the tray with the ice; steam swirls above the roiling liquid. In the early 1800s, in the first indication of the diversity of phase transitions, physicists found they could compress a gas into liquid without any abrupt change or transient co-existence. Later that century, physicists discovered that a bar of iron could undergo an analogous transformation from magnet to nonmagnet. In the twentieth century, they noticed phases within phases: water ice can switch from one crystal structure to perhaps two dozen others. And they discovered superconductivity, later recognized to be a phase not of the atoms but of the electrons within a material. Superconductivity—in which electrons move through a material without resistance—and the related phenomenon of superfluidity—in which liquids flow without friction—were the first known phases to depend in an essential way on quantum physics, which describes the counterintuitive physical properties that matter can exhibit when conditions deviate from those of everyday life.

Like the materials they study, physicists themselves went through a change of state in the 1960s. Recognizing that diverse forms of matter obey common principles of organization, researchers combined their disparate areas of study into a single subject: condensed-matter physics. Closely allied with chemistry and material science, but also exchanging ideas with atomic and particle physics, condensed matter is now the largest subdiscipline of physics, and its discoveries radiate into every field of science and technology. Within this broad area, the history of studying phase transitions is fascinating in its own right (Domb, 1996; Kadanoff, 2013; Blundell, 2019).

Today, there is an ever-growing set of candidate phases of matter. They fall into multiple broad categories, within each of which there are hundreds of variations depending on, for example, the symmetry of the arrangement of atoms in the matter (Wen, 2017; Song et al., 2020) or the computational complexity of their state (Liu et al., 2019a, b).

This review will certainly not attempt to cover all those phases, let alone the many other examples of emergence in condensed matter. Rather, in the remainder of this introductory section, it shall describe physicists' previous best understanding of phase transitions. Then it shall focus on three broad types of phases that in recent years have shown the most potential for furthering our foundational understanding of physics or for practical applications, such as for producing nanodevices or for implementing quantum computing. The three broad phases are: soft matter—including polymers and organic LEDs—which will be described in Sect. 2.2; quantum phases of matter—including superconductors, Bose–Einstein condensates, and strange metals—discussed in Sect. 2.3; and topological phases of matter—including the quantum Hall effect and topological insulators—covered in Sect. 2.4. Finally, Sect. 2.5 will look at a promising attempt to use insights from condensed matter to unify physics. Called the "string-net-liquid" model, it proposes that space itself takes the form of a peculiar kind of quantum fluid and that matter and light may be emergent from this liquid.

Table 2.1 lists some of the most prominent phases of matter that have been identified. These are rough categories only and the distinctions among them can be blurry. Some solids have liquid-like behavior, and some liquids act like solids. Classical (that is, non-quantum), quantum, and topological phases also overlap. These complications keep the subject interesting.

2.1.1 Just a Phase They're Going Through—Landau's Theory of Phase Transitions

Exotic phases of matter display genuinely new emergent properties that could not have been predicted by looking at the arrangement of their constituent atoms, alone. By tracking a material through a phase transition, theorists and experimentalists can see how atoms gain collectively what they lack individually.

States of matter arise from a competition between forces of order and forces of disorder. Those you see in a kitchen—ice cubes, running water, and puffs of steam—are differentiated by temperature. At low temperature, order

Table 2.1 Phases of matter

Type	Phase	Description
Classical	Solid	Retains its shape and volume. Atoms and molecules interact strongly to hold themselves in place. Could be either crystalline or amorphous (glassy)
	Liquid	Retains its volume, but lacks shear strength and acquires the shape of its container. Atoms and molecules interact strongly, yet remain disordered. A glass can be considered a very slowly flowing liquid
	Gas	Expands to fill the available volume. Atoms and molecules interact weakly, if at all
	Plasma	A gas of charged particles. Responds to electrical and magnetic as well as mechanical forces
	Soft Matter	Gels and polymers have properties of both solid and liquid. Some also resemble the topological phases mentioned below
Magnetic	Ferromagnet	A permanent magnet. Particle spins are ordered
	Paramagnet	Can be temporarily magnetized by an external field. Particle spins are disordered
	Antiferromagnet	Particle spins are ordered, but negate one another
Quantum	Metal	Conducts electricity and heat, albeit with some resistance. Though solid from the outside, internally it is a gas of electrons
	Mott Insulator	Electrical insulator in which electrons collectively seize up, like a subatomic traffic jam
	Superconductor	Conducts electricity without resistance. Usually but not always very cold. In the standard variety, electrons pair up and flow through the crystal as a liquid rather than as a gas
	Superfluid	Very cold liquid that flows without friction. It climbs out of containers and, if stirred, swirls forever.

(continued)

Table 2.1 (continued)

Type	Phase	Description
	Strange metal	Electrical resistance is directly proportional to temperature, indicating that electrons behave in a highly collective manner
Topological	Quantum hall effect (integer and fractional)	Voltage develops across the material in response to magnetic field. Electrical resistance increases in discrete steps (it is quantized). Electrons form complex choreographies
	Quantum spin liquid	Highly entangled but disordered system with peculiar magnetic properties
	Topological insulator	An electrical insulator in its interior, but a conductor along its outer surface. Not as strongly entangled as other systems defined by their topology
	String-net liquid	Hypothesized phase in which particle spins form extended structures
Non-equilibrium	Many-body localization	A poorly understood phase that resists the usual tendency to become disordered. One version becomes a perfect electrical insulator at low temperatures
	Time crystal	System that stably cycles through configurations

rules. Atoms or molecules settle into neat rows or other geometric patterns—a crystal—and their mutual electrical forces swiftly bring any deviants into line. As the temperature rises, thermal vibrations rattle this tidy arrangement and ultimately break it apart, whereupon the atoms might find some other arrangement or disperse altogether.

As a first peek inside the black box of phase transitions, physicists attempted to take into account these changes. They developed so-called mean-field theory—instead of tracking each and every interaction among the particles, they considered an averaged interaction. In the 1930s, the renowned Russian physicist Lev Landau developed such a theory that applied to all phase transitions, be it liquid water turning to steam or iron becoming demagnetized. Landau set aside the details and focused on the macroscopic properties that define a phase, such as density (for a fluid) or magnetization (for a magnet)—these are called "order parameters." He connected these order parameters to the concept of symmetry. The hotter phase has the greater symmetry. Water molecules in steam dart every which way, but water molecules in a snowflake line up in certain directions—thus the crystal is less symmetrical (and more beautiful for it). In the hotter phase, the order parameter is zero, and as the material cools, the order parameter rises to a value that defines the asymmetry of the new phase.

Because it works for systems as diverse as steam and magnets, Landau's theory captures an essential insight into phase transitions. They are collective phenomena. It is what the building blocks do, rather than what they are, that makes the difference. In a principle known as universality, systems made of vastly different ingredients behave the same as long as their broad-brush features, such as spatial dimensionality, are the same.

2.1.2 Tipping the Scales

Normally physics is segregated by scale, and thank goodness, because it makes the universe comprehensible to us. A doctor can treat your headache without worrying about how particles interact or how the universe expands. But certain kinds of phase transitions require coordination between the large and the small scales.

To see how this plays out, consider magnetism. Particles such as electrons carry an intrinsic magnetism, called "spin," which can point in one of two directions, up or down. In a bar magnet, all particle spins are initially aligned. But, as you warm up the bar magnet, the particles gradually all fall out of alignment until the iron is no longer magnetized. Physicists describe this coordinated behavior in terms of what they call the correlation length—the distance over which parts of the material interact.

Normally the correlation length is short. But the following sections will describe instances where it is surprisingly long, spanning entire chunks of material and leading to novel emergent phenomena. These complex patterns stretch across what is, to a physicist, an enormous distance. You can't see the patterns by cranking up the magnification on your microscope. An atom, in isolation, might or might not be part of a broader configuration. If anything, you need to pull back. This is especially true for the exotic phases described in Sects. 2.3 and 2.4—quantum and topological phases—which are produced in the lab under highly specialized conditions, often at frigid temperatures. The aforementioned renormalization group provides a mathematical technique to zoom in and out.

The standard theories of phase transition make various idealizations. Among other things, they imply that a system has to contain an infinite number of particles to undergo a phase transition. The reason is that phase transitions involve a discontinuity (although the nature of the discontinuity depends on the type of transition) and only an infinite system can undergo a genuinely abrupt change. Treating systems as infinite is a routine maneuver in physics, and the special mathematics of infinity can lead to strong emergence: Infinitely big systems will have collective properties that cannot be derived from the microscopic physics, but must be specified independently (Anderson, 1972; Gu et al., 2009).

Yet nothing in nature is truly infinite. Are the phase transitions we see illusory, then? In practice, the number of molecules is so large that, for these purposes, it might as well be infinite (Liu, 1999; Callender, 2001; Butterfield, 2011). Physicists in the 1970s and 80s developed a formalism to relate realistic systems to the infinite ideal (Fisher & Berker, 1982). More recently, they have found that, in quantum systems, the number of molecules need not be infinite, as long as the number of possible states is (Kais & Serra, 2003). Even a single atom has been observed to undergo a phase transition (Cai et al., 2021).

2.1.3 Non-equilibrium Phases of Matter

The phases that will be described in detail in this review, different though they are, have one feature in common: they occur in equilibrium. The systems are able to strike a balance between order and disorder. But it is worth mentioning before moving forward that even that requirement is now up in the air. Table 2.1 includes the broad category of non-equilibrium phases—transitions that occur in an unsteady condition that should be inimical to the development of new forms of order. These will not be covered in depth in this review, but have been subject of much recent investigation.

One example is the many-body localized phase (Bauer & Nayak, 2013; Nandkishore & Huse, 2015; Alet & Laflorencie, 2018; Altman, 2018; Abanin et al., 2019). In this peculiar state, particles stay put rather than disperse, as they are ordinarily wont to. The basic idea goes back to Philip Anderson in the 1950s and won him a Nobel Prize, but his analysis applied only to single particles; about 15 years ago, physicists began to extend the idea to vast systems of particles. These systems retain a memory of how they formed, which equilibrium systems lose; information does not smear out, as it does in equilibrium. For this reason, some have suggested using many-body localization as memory for a quantum computer. Another characteristic property of the phase is that the system remains an electrical insulator even when it should conduct. Physicists may even have observed systems transition from an equilibrium state into a many-body localized state (Ovadia et al., 2015).

A second out-of-equilibrium phase is the time crystal, which is defined not by the spatial arrangement of atoms or molecules, but by temporal regularity.

The system spontaneously oscillates even in its lowest-energy state. The original concept presumed thermal equilibrium, but later work showed that the system must be in disequilibrium. Chains of ions of particle spins can cycle through patterns in a stable and robust way, even though they are buffeted by external perturbations that, by rights, should throw off their rhythm (Wilczek, 2012; Choi et al., 2017; Zhang et al., 2017; Sacha & Zakrzewski, 2018).

2.2   Soft Matter

Household materials such as plastic, rubber, gels, foams, and even the tissue of your body may at first glance appear to be quite unlike one another. Yet, surprisingly, they share some counterintuitive properties. A new field of soft matter research has coalesced in the past two decades as physicists have come to recognize that these disparate materials are governed by common principles. Studying the flow and thermal behavior of soft matter is important for a wide variety of new technological applications, notably organic electronics—such as the organic LEDs (OLEDs) that are now common in computer displays and smartphone screens (Liu et al., 2015)—and tissue engineering—such as cancer therapy and organ regeneration (Gonzalez-Rodriguez et al., 2012).

The macroscopic properties of many materials can be understood in full by looking at their microstructure—at the crystalline arrangement of their atoms and molecules, for example. This is not true for soft matter, however. One of the characteristic features of emergent phases is that their macroscopic behavior cannot be predicted directly by looking at the arrangements of their atoms; rather you must zoom out to understand the origin of their order. In the case of soft matter, its macroscopic behavior emerges from the interactions among structures at an intermediate or "mesoscopic" scale. For instance, foam resists compression because the bubbles act as tiny springs (Cox, 2013). The bubbles are much larger than the foam's constituent atoms, but much smaller than the foam taken as a whole.

Another defining feature of soft matter is that its complex behavior arises spontaneously, as a result of self-assembly and organization. Take rubber as an example. When you stretch a rubber band, the force you feel is not due to a microscopic interaction such as molecular bonding, at least not directly. Rather, you are fighting the natural tendency of the rubber to flop around. There are more ways for the polymer chains in the rubber to be floppy than to be straight. Physicists quantify this in terms of entropy, which counts the number of different ways microstates in a system can be re-arranged (in this case, the different configurations of the polymer chains) while still giving the same specific macrostate of the system (the rubber band as a whole). A floppy rubber band has a higher entropy than a straight band. (Entropy can also be defined in other subtly different, but related ways, as shown in the next chapter.)

Left to their own devices, polymer chains will wind up in one of those more numerous states, appearing from the outside to be floppy, rather than in a less-likely arrangement in which the rubber band is perfectly straight. This is purely a statistical effect, one that arises from the relative probability of being floppy or straight. An analogy is your messy desk. There is no "messiness force" that causes piles of paper to sprawl and used coffee cups to accumulate. Rather, tidiness is a special state of affairs; there are many more ways to be messy, so almost anything you do will tend to make the desk messier.

Another collective effect is topological as opposed to statistical. Because polymer chains can't cross, they block one another's motion, so the material behaves like a solid even though technically it does not undergo a phase transition. It remains in an internal state of stress that releases only very slowly. These topological effects are mathematically parallel to those in some of the exotic quantum phases that the next chapter will describe (McLeish et al., 2019).

Soft materials are important both conceptually and practically. They are commonplace, they illustrate in a simple way the general principles of emergence, and they are routinely used to make devices at room temperatures, where the effects of quantum physics can be ignored (McLeish, 2019). The next two sections describe advances that have been made by considering far weirder materials—with even stranger characteristics—that physicists and engineers hope to exploit for a range of new technologies.

2.3   Quantum Phases of Matter

Nearly everyone, at some point, will entrust their life to a quantum phase of matter. MRI scanners in hospitals depend on superconducting magnets. So do particle accelerators, wind turbines, and maglev trains. Physicists are still grappling with understanding the mechanisms behind superconductivity in its many guises. At the same time, they are discovering other new quantum phases that can be produced in materials such as graphene—first isolated in 2004 (Novoselov, 2004)—which may offer even greater benefits.

2.3.1 Quantum Physics

Under everyday conditions, we see but a few of the ways that atoms and particles can arrange themselves, because the more sophisticated arrangements needed for exotic phases to arise are spoiled by the random thermal vibrations that buffet atoms. Only when physicists dial down the temperature, reducing these thermal jiggles, do they see the full flowering of self-order, as the distinctive features of quantum physics make themselves felt.

One of those features is uncertainty. The more you zoom into the microscopic realm, the fuzzier it gets. You can't predict with confidence where you will find a particle, because until it is measured the particle exists in a range of positions. When probed, it could materialize anywhere in that range, with the outcome decided randomly. Likewise, the particle does not have a perfectly defined momentum until it is observed; it spans a range of values and, if forced to settle into one of those values, does so at random. To add to the fuzziness, the particle's position and momentum are yoked together. If the particle is confined to a sufficiently narrow range of positions, the range of its momentum widens. You thus cannot know with perfect accuracy both where the particle is and where it is going. This tradeoff was captured by Werner Heisenberg in his famous uncertainty principle. Other aspects of particles such as their energy and, in some cases, their very existence are also subject to uncertainty.

Another distinctive feature of quantum physics is known as entanglement, which harmonizes the properties of multiple particles across huge distances, in a way that has no analogue in ordinary physics. Chapter 3 will explain entanglement in greater depth. Although quantum effects are typically most distinctive at the microscopic level, they exist at every scale, and in quantum phases of matter they can control the characteristics of a large slab of material.

As noted above, states of matter reflect a tug-of-war between order and disorder, and new tendencies pulling in either direction manifest themselves in the quantum realm. Because of Heisenberg's uncertainty principle, atoms and particles are inherently unable to hold still; the more tightly they are confined in position, the more they will wriggle. When the temperature of a material is reduced, they tend to stay put. But as particles are now more confined, their quantum jiggling actually becomes all the more intense, and the system must strike some new balance. It can be driven to reorganize en masse, creating new phases of matter.

2.3.2   Frozen yet Mobile—Superconductors, Bose–Einstein Condensates, and Strange Metals

Superconductivity was discovered experimentally by Dutch physicist Heike Kamerlingh Onnes as far back as 1911, when he was investigating the electrical conductivity of mercury at cryogenic temperatures, although at the time its mechanism was not fully understood (and arguably still isn't).

Metals conduct electricity as electrons flow through them. Typically, these electrons meet with resistance. That can change if temperatures are low enough. Electrons usually repel each other, as do any particles with the same electric charge, but under frigid conditions—and with the aid of the crystalline lattice in which they are embedded—electrons can overcome this native aversion to clumping and pair up. In vast numbers, they settle into the same energy state, like a clone army, forming a phase known as a Bose–Einstein condensate. The electron pairs are able to flow freely in this state, giving rise to what today is termed low-temperature superconductivity.

High-temperature superconductors—"high" meaning above –200 °C—were discovered experimentally in the 1980s (Bednorz & Müller, 1986). These involve a different mechanism from their colder cousins, and trying to figure them out has led physicists to numerous other findings, including the discovery of so-called strange metals—materials whose resistance has an unexpected relationship to temperature (Keimer et al., 2015). When normal, un-strange metals are examined at extremely low temperatures, their electrical resistance increases with the value of its temperature squared and eventually plateaus. By contrast, the resistance of strange metals is directly proportional to temperature (Bruin et al., 2013) under these conditions and also to the external magnetic field (Hayes et al., 2016; Giraldo-Gallo et al., 2018). This behavior, as well as other measurements, indicate that strange metals cannot be thought of as an assemblage of particles or even of clumps of particles. The electrons must organize themselves into more complicated structures (Kaminski et al., 2003). (Chapter 4 will describe how the theory of strange metals has been improved by borrowing mathematical elements originally developed by considering quantum gravity and black holes.)

New simulations show that strange metals are an example of a counterintuitive quantum phase of matter known as spin liquids (Cha et al., 2020). In an ordinary magnet, particle spins are aligned and act in lockstep, like soldiers standing in precise ranks before a parade stand. In a spin liquid, particles are disordered, like hippies at Woodstock. And yet they still remain highly correlated, due to a kind of long-range quantum entanglement, enabling their behavior to remain linked despite their confused arrangement (Sachdev, 2012; Savary & Balents, 2017; Broholm et al., 2020).

2.3.3 Critical Symmetry

At so-called critical points, a material is poised between phases, and a system of particles is at its most delightfully complex, with orchestrated activity on every level. In quantum phase transitions, when you can adjust multiple factors such as temperature and magnetic field simultaneously, the critical "point" enlarges into an entire region. Among their many surprising properties, quantum-critical materials can be symmetrical in a way that its underlying system isn't.

These emergent symmetries invert the usual thinking in physics. Normally, physicists regard nature at its fundamental level as highly symmetric and think the messiness of ordinary matter spoils it—"squalid-state physics," particle theorist Murray Gell-Mann famously said. But here nature is asymmetric at its root and spontaneously develops a spontaneous symmetry, not just in spite of its messiness, but because of it.

In 2010 Radu Coldea and his colleagues cranked up a magnetic field to force a magnetic material, cobalt niobate, to undergo a phase transition. Just as the material was passing through the quantum critical point, electrons in the system formed clumps with energies of very specific values, matching the most complex symmetry pattern known to mathematics, E_8 (Coldea et al., 2010) (Fig. 2.1). Other symmetries related to patterns in gravitational and particle physics also can emerge (Zaanen & Beekman, 2012).

2.3.4 "Twistronics"—Graphene's Magic Angle

Physicists can drive a material through a phase transition by squeezing it, bringing a magnet close by, sprinkling in new chemical ingredients, or trapping atoms with crisscrossing laser beams. For instance, one type of insulator, known as a Mott insulator, can be turned into a conductor either through a conventional thermal transition (effected by lowering the temperature) or through a quantum transition (precipitated by increasing the pressure or altering the chemical composition) (Guguchia et al., 2019). An insulator can also morph into a superconductor, using an electric or magnetic field (Fisher, 1990) or in some cases into high-temperature superconductors by tweaking

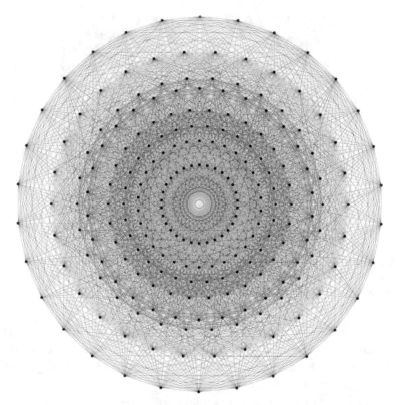

Fig. 2.1 E_8 is a complex 248-dimensional symmetry discovered in the late 1800s. The structure appears in calculations and patterns seen in many theories of physics (*Image credit* Jgmoxness shared under Creative Commons license CC BY-SA 3.0.) [https://com mons.wikimedia.org/wiki/File:E8Petrie.svg]

their chemical composition ever so slightly (Battisti et al., 2016). And a super-fluid can transform into an insulator (Orzel et al., 2001; Greiner et al., 2002; Vojta, 2003).

One tantalizing advance involves the much-studied graphene's ability to transition from a Mott insulator to a superconductor, simply by twisting it. Graphene is made from a single layer of carbon atoms linked in a hexagonal honeycomb pattern (Fig. 2.2). It excites engineers because it is exceptionally strong, conducts heat and electricity better than most metals, and can easily be chemically altered to produce transistors, sensors, and other components. In 2018, in one of the many surprises this field is capable of, Pablo Jarillo-Herrero and his team stacked two sheets of graphene with their grids offset by a "magic angle" of precisely 1.1 degrees. Just by making this slight twist, they caused the graphene layers to become superconducting (Cao et al., 2018a, b). When you stack the two hexagonal sheets on top of each other, larger

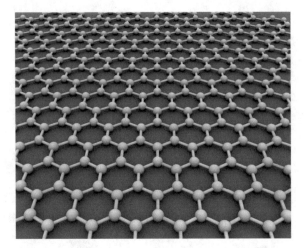

Fig. 2.2 Graphene is an atomic-scale hexagonal lattice made of carbon atoms (*Image credit* AlexanderAlUS, shared under Creative Commons license CC BY-SA 3.0.) [https://commons.wikimedia.org/wiki/File:Graphen.jpg]

hexagonal patterns begin to form, and these hexagons become the individual units, rather than the small hexagons traced out by the carbon atoms. The stacking thus lets electrons in each sheet become correlated and release one another to flow freely (Balents et al., 2020).

An electric field or mechanical pressure can force this sandwich, known as twisted bilayer graphene, to change from a superconductor to a Mott insulator (Yankowitz et al., 2019). The discovery has caused a flurry of activity, since graphene is easy to produce and this behavior is easy to control in the lab. A new field dubbed "twistronics" studies electronic behavior in twisted graphene and other materials, and a theory, based on geometric concepts, is beginning to emerge (Hu et al., 2019; Julku et al., 2020).

2.4   Topological Phases of Matter

The huge variety of phases of matter has been compared to a multitude of dance styles. As Wen (2013) puts it: Ordinary phases of matter are solo dances. A magnet is like the Electric Slide or Macarena: particles are all doing the same thing, but they've doing it individually. And a superconductor is like tango or salsa: particles dance in couples, but each couple is on its own.

In the late 1980s Wen realized that several newly discovered phenomena qualified as yet another kind of phase of matter: topological order (Wen,

1989; Zeng et al., 2019). Topological order is one of the most significant developments of recent decades. One example is the spin liquid—the frustrated magnets mentioned briefly above—which has a long-range order due to entanglement, even though its individual spins lack any alignment. Another is the quantum Hall effect, in which composite structures emerge that behave like particles carrying an amount of electric charge that is not otherwise found in nature. Topological concepts also define so-called topological insulators, which are materials that conduct electricity on their surface, while remaining electrically insulating in their interiors (Qi & Zhang, 2011). Topological materials are such a new find that their technological applications are still unclear. They offer the promise of more efficient devices based on spin-electronics or "spintronics," which, as the name suggests, would exploit the spin of electrons to create more efficient data storage and processing, as well as ways of quantum computing that lose less energy or are less prone to errors.

Topological phases eluded physicists for so long—and still elude simple explanation—because they are tricky to visualize. The topology in these cases is not to do with the physical shape of the system or the arrangement of its particles—patterns in space you could imagine viewing with a strong enough microscope. Rather, the topology involves the abstract geometry of the system's dynamics, creating patterns that coordinate how the particles are able to move over time. They vary in complexity depend on how much the particles interact with one another. This brings us back to dancing. To Wen, topological phases are crowd dances like square dancing or Cuban rueda, in which a caller coordinates the particles' actions to produce some grand pattern across the floor. In the quantum Hall effect, described more fully below, particles do-si-do around one another to the caller's instructions. Spin liquids are the most sophisticated choreographies of all. Particles form conga lines that weave in and out of one another, and these lines, rather than the dancers, are the building blocks of the larger pattern.

Such phases, with their complex coordinated dynamics, arise from the global structuring of a system (Wen, 2019). You might not even notice them if you look at a substance piecemeal. With topological insulators, only a global view reveals that an otherwise insulating material is able to conduct electricity (Hasan & Kane, 2010). Similarly, in topological superconductors, the bulk of the material is an insulator, yet the edge conducts electricity and heat without resistance (Beenakker & Kouwenhoven, 2016; Sato & Ando, 2017; Frolov et al., 2020). Further enriching the dynamics, a system can have phases of both the traditional kind that Landau identified, as described in Sect. 2.1.1, and the topological kind (González-Cuadra et al., 2019).

2.4.1   The Hall Effects

In the 1980s, physicists discovered a pivotal case study of emergence: the integer and fractional quantum Hall effects. Like Bose–Einstein condensates, described in Sect. 2.3.2, they demonstrate that entire blocks of material can dance to the quantum drumbeat. Their signature feature is that the electrical resistance of the material jumps in steps, rather than changing continuously, in the presence of magnetic fields. What is particularly tantalizing to physicists hoping to build new devices is that temperature fluctuations or impurities that might usually affect this resistance have no effect.

The Hall effect comes in three progressively more sophisticated versions (Tong, 2016). The classical version—which does not involve quantum physics—was discovered in the late nineteenth century by Edwin H. Hall. Let's say you have a long, thin and flat slab of an electrically conducting material, with a current flowing along its length (Fig. 2.3a). If you hold a magnet near the conductor, the magnetic field bends the current toward the side of the conductor. Negative electric charges build up along one long edge, leaving the opposite edge with a net positive charge, until the accumulating charges counteract the magnet's influence. These charges create a transverse voltage difference across the width of the slab (Fig. 2.3b). This measurable transverse voltage is called the Hall effect.

This behavior creates a variation on electricians' Ohm's law, the relation between current, voltage, and electrical resistance. The resistance along the path of the current is very low, perhaps even zero, but a considerable resistance can build up in the transverse direction, and this resistance increases with the magnetic field (Fig. 2.3c). Smartphones carry a magnetic sensor that exploits this effect to detect the phone's orientation.

Now, if you cool the conductor to the temperature of liquid helium (just around 4 °C above absolute zero), the integer quantum version of the Hall effect awakens. Then the resistance does not increase steadily with the magnetic field, but walks up a staircase (Klitzing et al., 1980; Thouless et al., 1982). This happens because the magnetic field deflects the electrons so strongly that they travel in closed loops (Fig. 2.3d). Quantum effects limit this circular motion to regularly spaced frequencies, which restricts the energy that the electrons can have and, together with other effects, creates discrete steps in the material's traverse electrical resistance. Because the electrical resistance no longer increases smoothly with magnetic field, but grows in jumps, it is said to be "quantized."

Physicists are relatively used to seeing quantized behavior in the microrealm. Quantum physics earned its name because it describes, among other things, how the energy states of atoms can only take discrete values. What is so surprising in the quantum Hall effect is that physicists are observing quantization in a macroscopic property, resistance. This is remarkable because electrical resistance is ordinarily a very messy phenomenon. Billions of flowing electrons have a complicated herd behavior that varies with temperature, magnetic field, and chemical impurities. Yet all those complications go away in the quantum Hall effect. The resistance is extremely precise and reproducible. That is a sign that the effect is ultimately due to a genuinely new effect, topology, which, by definition, does not depend on precise geometric details. Early experiments used gold leaf as the wire, while later ones used transistors of silicon or gallium arsenide. Even more recently, physicists have seen the effect in graphene (Novoselov et al., 2007; Bolotin et al., 2009).

As you crank up the magnetic field, additional quantized levels occur, as if the stairs were shrinking to midget size (Fig. 2.3e) (Tsui et al. 1982). Before, the electrons behaved almost independently. But now, in this new phase of matter, they interact with one another and clump together in an extremely peculiar fashion. The system behaves as though it is made up not of electrons, but of emergent "quasiparticles" with a fraction of the electron's charge and a value of spin that no electron or indeed any other elementary particle carries (Laughlin, 1983; Wen & Niu, 1990). This is the fractional quantum Hall effect. It is genuinely bizarre given that electrons are themselves indivisible particles, yet the quasiparticles that emerge can have charges of a 1/3 or a 1/5 of an electron. Fascinating its own right, the fractional quantum Hall also reveals a broader class of systems called topological quantum liquids (Swingle & McGreevy, 2016).

There is added point of irony here. Experimentally, the quantum Hall staircase is not only immune to messiness, but demands it. It will only appear in a disordered or dirty sample. As the sample becomes purer, physicists measure more and more stairs at intermediate fractions. But in the limit, a line with an infinite number of tiny steps reduces to a straight line, which is precisely the classical result for Hall conductivity.

Multiple Nobel prizes were awarded for the experimental discovery and theoretical explanation of the quantum Hall effects. They laid the foundation for all future discoveries of topological order. Until the mid-2000s, the effects had been induced by physicists using extremely strong magnetic fields. But physicists were about to discover that some materials could create their own complex electron dynamics, thanks to the magnetic interactions between their electrons and atomic nuclei. Electrons are able to flow almost without resistance on the surface of these materials, which are called topological insulators, because their interiors are electrical insulators.

Hall Effects

The Hall effect can come in classical or quantum forms. The quantum version is an emergent property of vast numbers of electrons, which underlies the phenomena seen in many topological materials, such as topological insulators.

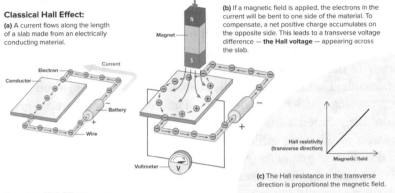

Classical Hall Effect:

(a) A current flows along the length of a slab made from an electrically conducting material.

(b) If a magnetic field is applied, the electrons in the current will be bent to one side of the material. To compensate, a net positive charge accumulates on the opposite side. This leads to a transverse voltage difference — **the Hall voltage** — appearing across the slab.

(c) The Hall resistance in the transverse direction is proportional the magnetic field.

Quantum Hall Effect:

(d) Electrons within the slab are deflected so strongly that, in effect, they are confined to moving in closed in loops. No current flows within the bulk of the slab. But electrons can hop along the edge of the material, conducting electricity, almost without resistance.

Quantum Spin Hall Effect:

(f) In some materials, each electron's spin can become coupled to its motion — even in the absence of an external magnetic field, due to internal magnetic effects. Electrons that are spin 'up' will flow around the edge in one sense, while those that are spin 'down' will travel in the opposite sense. Such effects are extremely robust, making them ideal candidates for new types of electronic devices that are fast, energy efficient and resilient against data corruption.

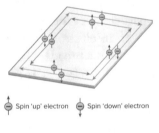

Spin 'up' electron   Spin 'down' electron

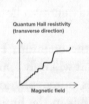

(e) The circling motion restricts the energies that electrons can have to discrete values. In turn, the quantum Hall resistance rises in discrete steps as the magnetic field increases. This is the integer quantum Hall effect. For very strong magnetic fields, electrons form collective structures with a fraction of the charge of an electron. This is the fractional quantum Hall effect.

2.4.2 Topological Insulators and Topological Nanomaterials

Kane and Mele (2005) predicted topological insulators as a byproduct of looking for the quantum Hall effect in graphene. Within three years experimentalists had synthesized them in the lab (Konig et al., 2007; Hsieh et al., 2008). As mentioned, these behave as insulators on the inside, while conducting electricity along their surface.

◄**Fig. 2.3** Hall effects. The Hall effect can come in classical or quantum forms. The quantum version is an emergent property of vast numbers of electrons, which underlies the phenomena seen in many topological materials, such as topological insulators. **a** A current flows along the length of a slab made from an electrically conducting material. **b** If a magnetic field is applied, the electrons in the current will be bent to one side of the material. To compensate, a net positive charge accumulates on the opposite side. This leads to a transverse potential difference—the Hall voltage—appearing across the slab. **c** The Hall resistance in the transverse direction is proportional to the magnetic field. **d** Electrons within the slab are deflected so strongly that, in effect, they are confined to moving in closed in loops. No current flows within the bulk of the slab. But electrons can hop along the edge of the material, conducting electricity, almost without resistance. **e** The circling motion restricts the energies that electrons can have to discrete values. In turn, the quantum Hall resistance rises in discrete steps as the magnetic field increases. This is the integer quantum Hall effect. For very strong magnetic fields, electrons form collective structures with a fraction of the charge of an electron. This is the fractional quantum Hall effect. **f** In some materials, each electron's spin can become coupled to its motion—even in the absence of an external magnetic field, due to internal magnetic effects. Electrons whose spin points up will flow around the edge in one sense, while those whose spins point down will travel in the opposite sense. Such materials are extremely robust, making them ideal candidates for new types of electronic devices that are fast, energy efficient, and resilient against data corruption (*Image credit* Created by Elfy Chiang for the Foundational Questions Institute, FQXi. © FQXi)

The difference between the interior and surface was already evident in the quantum Hall effects (Moore, 2010; Maciejko et al., 2011; Qi & Zhang, 2011; Wang & Zhang, 2017). Under the influence of the magnetic field, electrons can be thought of as moving in little circles in the bulk. At the material's edge, however, electrons do not have the space to complete full circles; instead they hop along in semi-circular jumps. Consequently, a current can develop along the length of the material at its edge, but not in its interior. By symmetry, the current flows in opposite ways along the two sides of the material. Because the two directions of current are neatly separated, like lanes on a motorway, electrons do not get in one another's way and can flow without resistance. This behavior is topological in the sense that it depends only on the most basic geometric property of the material: that it has two sides. The precise shape of those sides is unimportant.

Things get even more interesting when a material is engineered so that each electron's spin is coupled to its motion. Then the currents along the two sides differ not just in motion, but in spin; spin-up electrons move in one sense, while spin-down electrons move in the opposite sense (Fig. 2.3f). This so-called quantum spin Hall effect needs no imposed magnetic field and is even more robust than before; mathematically, it corresponds to a higher-order topology. Although the effect is restricted to a slab of material so thin that it is effectively two-dimensional, physicists soon found other insulators that are fully three-dimensional, forming a new class of materials known as topological insulators.

The sensitivity to spin suggests that these insulators might allow electronics based on electrons' spin rather than their electric charge (He et al., 2019; Tokura et al. 2019). It takes less energy to rotate a spin and thus control a spin current than it does to manipulate an electronic charge, making them ideal for creating faster and more energy-efficient electronic devices. Topology, as described above for the quantum Hall effect, also makes the system more resilient to data corruption because interactions with the environment, such as the exchange of heat, have little to no disruptive effects.

For realistic applications in devices, however, physicists must be able to control and detect topological states in materials by engineering them at the atomic level. Some groups are keenly investigating nanomaterials—structures, just tens to hundreds of nanometers across, that form by condensing out of vapor like dewdrops on blades of grass. Superconducting topological nanowires, for instance, are good candidates for robust quantum computing. The question remains, though, how stable the topological patterns are (McGinley & Cooper, 2020; Liu et al. 2019a, b).

2.4.3   Hunting for New Topological Materials—Can A.I. Help?

By 2019 physicists had found hundreds of topological materials and studied around a dozen in detail. In that year, the pace of discovery quickened even further. When a team of researchers scanned their databases to see whether any known materials might harbor topological states that had been overlooked, they found that almost a quarter of the materials were potentially topological—stacking up to thousands, in total (Zhang et al., 2019). The roster of candidates grows almost by the month (Choudhary et al., 2020). But how do you set about searching for topological states, when physicists do not know all the myriad forms they may take?

Enter machine learning. Having revolutionized image classification and language translation, machine learning is now doing the same for condensed-matter physics. In 2017 Giuseppe Carleo and Matthias Troyer analyzed the behavior of theoretical one- and two-dimensional grids of electrons using a neural network (Carleo & Troyer, 2017). To represent this system, they designed their neural network to take a configuration of electron spins as its input and create various combinations of those spins to capture the correlations they could have, in principle. Of the astronomically many correlations that might have been meaningful, the network identifies the most important, revealing the essential physics within the system.

The network can then be tuned to match experimental data. Whereas conventional techniques, known as quantum state tomography, may need a million data points to reconstruct the workings of even a modest system, Carleo and Troyer's method might get by with just 100. The network can also be trained to recognize phases of matter, given lots of labeled configurations (van Nieuwenburg et al. 2017). What is remarkable is that it can make sense of configurations that it was never trained on. It evidently picks up on genuine patterns that characterize phases—patterns that might not be evident to a human eye. This is especially true for topological, non-equilibrium, and glassy phases, which have subtle long-range correlations that even experts can miss and are hard to classify by traditional methods (Carrasquilla & Melko, 2017; Venderley et al. 2018; Bapst et al., 2020).

Researchers are now seeking to combine neural networks with quantum computers, creating the world's ultimate classification systems (Biamonte et al., 2017; Carleo et al., 2019).

2.4.4 Entangled in More Ways Than One—Topological Quantum Computing

Classical physics is based on the same principle as Legos. Big things are built out of many small pieces and derive their properties from them. But the materials we have been discussing demonstrate two ways that a whole can be more than the sum of its parts (Teller, 1986; Humphreys, 1997; Silberstein & McGeever, 1999). First, the system can have topological properties. These are, by definition, global. They transcend their parts. Indeed, a new emergent unit can arise from the composite behavior of the fundamental physical particles. Second, quantum entanglement creates a holistic linkage among particles which similarly defies the Lego principle. Entanglement is invisible if you study parts of a system in isolation from one another; you can recognize its presence only when you consider those parts jointly and notice unusual correlations between them.

Topology and entanglement act as the choreographers of the elaborate dances seen in the new states of matter. Though distinct concepts, they are related in practice. Topological states are highly entangled, although that is not their defining property (Bain, 2019; McLeish et al., 2019). Entangled links span large distances within the material, which is how individual particles can be sensitive to their collective state (Kitaev & Preskill, 2006; Chen et al., 2010; Lancaster & Pexton, 2015; Wen, 2019).

It is this global sensitivity that has led some physicists to posit that topological phases could serve as qubits—the quantum version of computer bits—for future quantum computers (Nayak et al., 2008). One practical obstacle to building quantum computers has been that qubits tend to be delicate creatures, easily disrupted by noise from heat in their surroundings or even just from a physical knock to the apparatus. This can lead to corruption of data, if not crash the whole system.

Topological phases offer a way out. There is increasing evidence that quasiparticles in the fractional quantum Hall effect fall into a hypothesized class of particle called "anyons" which retain a partial memory of where they have been, as if they were trailing string behind them. When multiple anyons are moved across a material's surface, their paths interweave to form a braid. The proposal is that data can be encoded in the braid rather than in a single anyon. This makes the system more robust to noise. If any single anyon is disrupted, it does not matter, because information is topologically encoded in the braiding traced across the entire material, which remains intact.

Experimentalists have been able to create such particles and build a system that have the requisite properties (Camino et al., 2005; Willett et al., 2013, 2019). In 2020 James Nakamura and Michael Manfra reported the strongest evidence yet that fractional quantum Hall states display anyonic behavior—fractional statistics and the memory effect—in a flat, effectively two-dimensional material made of gallium arsenide, with multiple layers to shield the system from noise (Nakamura et al., 2020). The system is still not sophisticated enough to support topological quantum computing, however. Other teams, meanwhile, have sought candidate particles in other systems such as conventional superconductors (Mourik et al., 2012; Nadj-Perge et al., 2014).

It should be noted that the field of topological quantum computing is immature compared to other modes of quantum computing that are being investigated. Using conventional superconductor technology, Google, IBM, and other companies have built computers with dozens of qubits, although they remain error-prone. Another firm, D-Wave, has created quantum processors with thousands of qubits, but these machines are not yet considered true computers that can execute whatever instructions you give them. Although it is something of a minority pursuit, topological quantum computing's promise of robustness to errors makes it an interesting endeavor.

2.4.5   Topological Inspiration for Other Areas of Science

Hard though topological phases of matter are to visualize, the underlying idea has filtered into other areas of science. Consider a system that looks nothing like an electrical conductor: Earth's oceans. Certain types of waves ripple through the seas along the equator as though pegged to an invisible line. In an analysis that applied ideas from condensed matter, atmospheric scientists explained these waves topologically (Delplace et al., 2017). To understand what happens at the equator, the researchers did not directly study the equator, but higher latitudes. That seems like a step backward, but they showed that you don't need to put yourself at the equator to deduce what happens there.

Within an abstract space of waves of possible wavelengths and amplitudes, the equations break down at specific places known as singularities. That is a mathematical consequence of Earth's rotation, which creates a Coriolis force that switches direction at the equator. The equator acts as type of boundary like those in condensed-matter systems—not a literal wall, but a place where the dynamics undergoes a step change. In a topological analysis like this, details such as water buoyancy and continental geography fall away. It doesn't matter that Earth is not a perfect sphere; such waves would be seen on flattened or even hypothetical ring planets. Any planet that rotates, whatever its shape or composition, will support these waves.

In fact, the planet doesn't even need to rotate, if the fluid itself is self-propelled. A bird flock is an example of a self-propelled system. If a flock were large enough to feel the curvature of the planet, it would experience a force similar to the Coriolis force as well as equatorial waves (Shankar et al., 2017). This is probably not relevant to bird flocks, but the authors suggest that cells in the body move on curved surfaces, and the theory could describe waves that affect their function. In this way, the physics of exotic forms of matter could have a direct effect on daily life, reminding us that principles of emergence are ubiquitous in the natural world.

2.5   Condensed Matter and the Unification of Physics—The String-Net-Liquid Model

Quantum and topological phases of matter show us that clumps of physical particles can behave like particles in their own right, with fractional charges, seemingly impossible spin values, and strangely choreographed motions. That suggests a radical thought. Might the particles we now take to be elementary actually be emergent from complex quantum and topological effects (Levin & Wen, 2005)? Physicists already know that many of the particles they used to take as fundamental, such as the protons and neutrons that make up atomic nuclei, are clumps of finer particles called quarks. But might those quarks, too, be emergent? If so, particle physics might be a type of condensed matter, and an entire level of reality might await our discovery (Zhang, 2002; Bain, 2019; Hamma & Markopoulou, 2011).

Wen has been actively pursuing this idea since at least 2002 (Wen, 2002). His model describes space as a topological phase called a string-net liquid and posits how both elementary particles and light may emerge (Wen, 2018, 2019). (This should not be confused with string theory, which is discussed in the next chapter.) On this account, space is a system of spinning particles arrayed in a lattice on which structures such as loops and chains form. These particles are some new building blocks distinct from electrons, quarks, and the like. Their loops and chains are quantum-entangled and therefore represent a long-range structuring of the particles. In fact, the individual particles become irrelevant; all the interesting dynamics occur at the level of these emergent structures.

The strings move around and fluctuate in density, as if they made up a liquid—but no ordinary liquid. Because the strings are linear structures with an orientation in space, their fluctuations have a directionality, like the orientation of an electric or magnetic field. When waves ripple through the liquid, the string density varies at right angles to the direction of wave motion. This is immediately reminiscent of electromagnetic waves, such as light, which propagate as oscillating electric and magnetic fields, at right angles to one another. Wen and his co-authors assert that not only is this behavior "like" electromagnetism, it *is* the origin of electromagnetism.

When fluctuations are muted, only a few, small closed loops form; as they become more intense, longer structures proliferate. The closed loops give rise to particles of light, photons, while the ends of open chains correspond to electrons (Levin & Wen, 2005). Although disturbances in the grid travel arbitrarily fast, the fastest are suppressed, leading to a limit on signal propagation. This fits with the observation that the speed of light is a constant, which cannot be surpassed by anything. In this formulation—unlike that of standard physics—the speed of light is not simply a fundamental and arbitrary constant, but an emergent property. Wen and his co-authors have extended the string-net-liquid model to describe the emergence of other elementary particles, such as quarks (Wen, 2003) and gravitons (Gu & Wen, 2006, 2012).

Among the many proposals for a fundamental theory of light and matter, the string-net liquid is a long shot. But it exemplifies the disciplinary crossovers among condensed-matter physics, particle physics, and gravity theory. The next two chapters will do a deep-dive into those improbable connections.

References

Abanin, D. A., Altman, E., Bloch, I., & Serbyn, M. (2019). Colloquium : Many-body localization, thermalization, and entanglement. *Reviews of Modern Physics, 91*(2). https://doi.org/10.1103/revmodphys.91.021001

Alet, F., & Laflorencie, N. (2018). Many-body localization: An introduction and selected topics. *Comptes Rendus Physique, 19*(6), 498-525. https://doi.org/10.1016/j.crhy.2018.03.003

Altman, E. (2018). Many-body localization and quantum thermalization. *Nature Physics, 14*(10), 979-983. https://doi.org/10.1038/s41567-018-0305-7

Anderson, P. W. (1972). More Is Different. *Science, 177*(4), 393-396. https://doi.org/10.1126/science.177.4047.393

Bain, J. (2019). Non-locality in intrinsic topologically ordered systems. *Studies in History and Philosophy of Science Part B: Studies in History and Philosophy of Modern Physics* **66**, 24-33.

Balents, L., Dean, C. R., Efetov, D. K., & Young, A. F. (2020). Superconductivity and strong correlations in moiré flat bands. *Nature Physics, 16*(7), 725-733. https://doi.org/10.1038/s41567-020-0906-9

Bapst, V., Keck, T., Grabska-Barwińska, A., Donner, C., Cubuk, E. D., Schoenholz, S. S., Obika, A., Nelson, A. W. R., Back, T., Hassabis, D., & Kohli, P. (2020). Unveiling the predictive power of static structure in glassy systems. *Nature Physics, 16*(4), 448-454. https://doi.org/10.1038/s41567-020-0842-8

Battisti, I., Bastiaans, K. M., Fedoseev, V., de la Torre, A., Iliopoulos, N., Tamai, A., Hunter, E. C., Perry, R. S., Zaanen, J., Baumberger, F., & Allan, M. P. (2016). Universality of pseudogap and emergent order in lightly doped Mott insulators. *Nature Physics, 13*(1), 21-25. https://doi.org/10.1038/nphys3894

Bauer, B., & Nayak, C. (2013). Area laws in a many-body localized state and its implications for topological order. *Journal of Statistical Mechanics: Theory and Experiment, 2013*(09), P09005. https://doi.org/10.1088/1742-5468/2013/09/p09005

Bednorz, J. G., & Müller, K. A. (1986). Possible High T_c Superconductivity in the Ba-La-Cu-O System. *Zeitschrift für Physik B Condensed Matter, 64*(2), 189–193. https://doi.org/10.1007/bf01303701

Beenakker, C., & Kouwenhoven, L. (2016). A road to reality with topological superconductors. *Nature Physics, 12*(7), 618–621. https://doi.org/10.1038/nphys3778

Biamonte, J., Wittek, P., Pancotti, N., Rebentrost, P., Wiebe, N., & Lloyd, S. (2017). Quantum machine learning. *Nature, 549*(7671), 195–202. https://doi.org/10.1038/nature23474

Blundell, S. J. (2019). Phase Transitions, Broken Symmetry and the Renormalization Group. In *The Routledge Handbook of Emergence* (pp. 237–247). Routledge. https://doi.org/10.4324/9781315675213-20

Bolotin, K. I., Ghahari, F., Shulman, M. D., Stormer, H. L., & Kim, P. (2009). Observation of the fractional quantum Hall effect in graphene. *Nature, 462*(7270), 196–199. https://doi.org/10.1038/nature08582

Broholm, C., Cava, R. J., Kivelson, S. A., Nocera, D. G., Norman, M. R., & Senthil, T. (2020). Quantum spin liquids. *Science, 367*(6475), 227–231. https://doi.org/10.1126/science.aay0668

Bruin, J. A., Sakai, H., Perry, R. S., & Mackenzie, A. P. (2013). Similarity of scattering rates in metals showing T-linear resistivity. *Science, 339*(6121), 804–807. https://doi.org/10.1126/science.1227612

Butterfield, J. (2011). Less is Different: Emergence and Reduction Reconciled. *Foundations of Physics* **41**, 1065-1135.

Cai, M.-L., Liu, Z.-D., Zhao, W.-D., Wu, Y.-K., Mei, Q.-X., Jiang, Y., He, L., Zhang, X., Zhou, Z.-C., and Duan, L.-M. (2021). Observation of a quantum phase transition in the quantum Rabi model with a single trapped ion. *Nat Commun, 12*(1), 1126. https://doi.org/10.1038/s41467-021-21425-8

Callender, C. (2022). Taking Thermodynamics Too Seriously. *Studies In History and Philosophy of Science Part B* **32**, 539-553.

Camino, F. E., Zhou, W., & Goldman, V. J. (2005). Aharonov-Bohm superperiod in a Laughlin quasiparticle interferometer. *Phys Rev Lett, 95*(24), 246802. https://doi.org/10.1103/PhysRevLett.95.246802

Cao, Y., Fatemi, V., Demir, A., Fang, S., Tomarken, S. L., Luo, J. Y., Sanchez-Yamagishi, J. D., Watanabe, K., Taniguchi, T., Kaxiras, E., Ashoori, R. C., & Jarillo-Herrero, P. (2018a). Correlated insulator behavior at half-filling in magic-angle graphene superlattices. *Nature, 556*(7699), 80-84. https://doi.org/10.1038/nature26154

Cao, Y., Fatemi, V., Fang, S., Watanabe, K., Taniguchi, T., Kaxiras, E., & Jarillo-Herrero, P. (2018b). Unconventional superconductivity in magic-angle graphene superlattices. *Nature, 556*(7699), 43-50. https://doi.org/10.1038/nature26160

Carleo, G., Cirac, I., Cranmer, K., Daudet, L., Schuld, M., Tishby, N., Vogt-Maranto, L., & Zdeborová, L. (2019). Machine learning and the physical sciences. *arXiv.org, physics.comp-ph.* https://arxiv.org/abs/1903.10563v1

Carleo, G., & Troyer, M. (2017). Solving the quantum many-body problem with artificial neural networks. *Science, 355*(6325), 602-606. https://doi.org/10.1126/science.aag2302

Carrasquilla, J., & Melko, R. G. (2017). Machine learning phases of matter. *Nature Physics, 13*(5), 431-434. https://doi.org/10.1038/nphys4035

Cha, P., Wentzell, N., Parcollet, O., Georges, A., & Kim, E.-A. (2020). Linear resistivity and Sachdev-Ye-Kitaev (SYK) spin liquid behavior in a quantum critical metal with spin-1/2 fermions. *Proceedings of the National Academy of Sciences*, 202003179. https://doi.org/10.1073/pnas.2003179117

Chen, X., Gu, Z.-C., & Wen, X.-G. (2010). Local unitary transformation, long-range quantum entanglement, wave function renormalization, and topological order. *Physical Review B, 82*(15). https://doi.org/10.1103/physrevb.82.155138

Choi, S., Choi, J., Landig, R., Kucsko, G., Zhou, H., Isoya, J., Jelezko, F., Onoda, S., Sumiya, H., Khemani, V., von Keyserlingk, C., Yao, N. Y., Demler, E., & Lukin, M. D. (2017). Observation of discrete time-crystalline order in a disordered dipolar many-body system. *Nature, 543*(7644), 221-225. https://doi.org/10.1038/nature21426

Choudhary, K., Garrity, K. F., Jiang, J., Pachter, R., & Tavazza, F. (2020). Computational search for magnetic and non-magnetic 2D topological materials using unified spin–orbit spillage screening. *npj Computational Materials, 6*(1). https://doi.org/10.1038/s41524-020-0319-4

Coldea, R., Tennant, D. A., Wheeler, E. M., Wawrzynska, E., Prabhakaran, D., Telling, M., Habicht, K., Smeibidl, P., & Kiefer, K. (2010). Quantum criticality in an Ising chain: experimental evidence for emergent E_8 symmetry. *Science, 327*(5962), 177-180. https://doi.org/10.1126/science.1180085

Cox, S. (Ed.). (2013). *Foams: Structure and Dynamics*. Oxford University Press. https://doi.org/10.1093/acprof:oso/9780199662890.001.0001

Delplace, P., Marston, J. B. & Venaille, A. (2017). Topological origin of equatorial waves. *Science* **358**, 1075-1077.

Domb, C. (1996). *The Critical Point*. Taylor & Francis. https://doi.org/10.4324/9780203211052

Fisher, M. E., & Berker, A. N. (1982). Scaling for first-order phase transitions in thermodynamic and finite systems. *Phys. Rev. B, 26*(5), 2507-2513. https://doi.org/10.1103/physrevb.26.2507

Fisher, M. P. A. (1990). Quantum phase transitions in disordered two-dimensional superconductors. *Physical Review Letters, 65*(7), 923-926. https://doi.org/10.1103/physrevlett.65.923

Frolov, S. M., Manfra, M. J., & Sau, J. D. (2020). Topological superconductivity in hybrid devices. *Nature Physics, 16*(7), 718-724. https://doi.org/10.1038/s41567-020-0925-6

Giraldo-Gallo, P., Galvis, J. A., Stegen, Z., Modic, K. A., Balakirev, F. F., Betts, J. B., Lian, X., Moir, C., Riggs, S. C., Wu, J., Bollinger, A. T., He, X., Božović, I., Ramshaw, B. J., McDonald, R. D., Boebinger, G. S., & Shekhter, A. (2018). Scale-invariant magnetoresistance in a cuprate superconductor. *Science, 361*(6401), 479-481. https://doi.org/10.1126/science.aan3178

González-Cuadra, D., Bermudez, A., Grzybowski, P. R., Lewenstein, M., & Dauphin, A. (2019). Intertwined topological phases induced by emergent symmetry protection. *Nat Commun, 10*(1), 2694. https://doi.org/10.1038/s41467-019-10796-8

Gonzalez-Rodriguez, D., Guevorkian, K., Douezan, S., & Brochard-Wyart, F. (2012). Soft matter models of developing tissues and tumors. *Science, 338*(6109), 910–917. https://science.sciencemag.org/content/338/6109/910.abstract

Greiner, M., Mandel, O., Esslinger, T., Hänsch, T. W., & Bloch, I. (2002). Quantum phase transition from a superfluid to a Mott insulator in a gas of ultracold atoms. *Nature, 415*(6867), 39-44. https://doi.org/10.1038/415039a

Gu, M., Weedbrook, C., Perales, Á., & Nielsen, M. A. (2009). More really is different. *Physica D: Nonlinear Phenomena, 238*(9-10), 835-839. https://doi.org/10.1016/j.physd.2008.12.016

Gu, Z.-C., & Wen, X.-G. (2006). A lattice bosonic model as a quantum theory of gravity. *arXiv.org, gr-qc.* https://arxiv.org/abs/gr-qc/0606100v1

Gu, Z.-C., & Wen, X.-G. (2012). Emergence of helicity ±2 modes (gravitons) from qubit models. *Nuclear Physics B, 863*(1), 90-129. https://doi.org/10.1016/j.nuclphysb.2012.05.010

Guguchia, Z., Frandsen, B. A., Santos-Cottin, D., Cheung, S. C., Gong, Z., Sheng, Q., Yamakawa, K., Hallas, A. M., Wilson, M. N., Cai, Y., Beare, J., Khasanov, R., De Renzi, R., Luke, G. M., Shamoto, S., Gauzzi, A., Klein, Y., & Uemura, Y. J. (2019). Probing the quantum phase transition in Mott insulator $BaCoS_2$ tuned by pressure and Ni substitution. *Physical Review Materials, 3*(4). https://doi.org/10.1103/physrevmaterials.3.045001

Hamma, A., & Markopoulou, F. (2011). Background-independent condensed matter models for quantum gravity. *New Journal of Physics, 13*(9), 095006. https://doi.org/10.1088/1367-2630/13/9/095006

Hasan, M. Z., & Kane, C. L. (2010). Colloquium: Topological insulators. *Reviews of Modern Physics, 82*(4), 3045-3067. https://doi.org/10.1103/revmodphys.82.3045

Hayes, I. M., McDonald, R. D., Breznay, N. P., Helm, T., Moll, P. J. W., Wartenbe, M., Shekhter, A., & Analytis, J. G. (2016). Scaling between magnetic field and temperature in the high-temperature superconductor $BaFe_2(As_{1-x}P_x)_2$. *Nature Physics, 12*(10), 916-919. https://doi.org/10.1038/nphys3773

He, M., Sun, H., & He, Q. L. (2019). Topological insulator: Spintronics and quantum computations. *Frontiers of Physics, 14*(4). https://doi.org/10.1007/s11467-019-0893-4

Hsieh, D., Qian, D., Wray, L., Xia, Y., Hor, Y. S., Cava, R. J., & Hasan, M. Z. (2008). A topological Dirac insulator in a quantum spin Hall phase. *Nature, 452*(7190), 970-974. https://doi.org/10.1038/nature06843

Hu, X., Hyart, T., Pikulin, D. I., & Rossi, E. (2019). Geometric and Conventional Contribution to the Superfluid Weight in Twisted Bilayer Graphene. *Phys Rev Lett, 123*(23), 237002. https://doi.org/10.1103/PhysRevLett.123.237002

Humphreys, P. (1997). How Properties Emerge. *Philosophy of Science, 64*(1), 1-17. https://doi.org/10.1086/392533

Julku, A., Peltonen, T. J., Liang, L., Heikkilä, T. T., & Törmä, P. (2020). Superfluid weight and Berezinskii-Kosterlitz-Thouless transition temperature of twisted bilayer graphene. *Physical Review B*, *101*(6). https://doi.org/10.1103/physrevb. 101.060505

Kadanoff, L. P. (2013). Theories of Matter: Infinities and Renormalization. In *The Oxford Handbook of the Philosophy of Physics*. Oxford University Press. https://doi.org/10.1093/oxfordhb/9780195392043.013.0005

Kais, S., & Serra, P. (2003). Finite-Size Scaling for Atomic and Molecular Systems. In (pp. 1–99). John Wiley & Sons, Inc. https://doi.org/10.1002/0471428027.ch1

Kaminski, A., Rosenkranz, S., Fretwell, H. M., Li, Z. Z., Raffy, H., Randeria, M., Norman, M. R., & Campuzano, J. C. (2003). Crossover from coherent to incoherent electronic excitations in the normal state of $Bi_2Sr_2CaCu_2O_{8+\delta}$. *Phys Rev Lett*, *90*(20), 207003. https://doi.org/10.1103/PhysRevLett.90.207003

Kane, C. L., & Mele, E. J. (2005). Z_2 topological order and the quantum spin Hall effect. *Phys Rev Lett*, *95*(14), 146802. https://doi.org/10.1103/PhysRevLett.95.146802

Keimer, B., Kivelson, S. A., Norman, M. R., Uchida, S., & Zaanen, J. (2015). From quantum matter to high-temperature superconductivity in copper oxides. *Nature*, *518*(7538), 179-186. https://doi.org/10.1038/nature14165

Kitaev, A., & Preskill, J. (2006). Topological entanglement entropy. *Phys Rev Lett*, *96*(11), 110404. https://doi.org/10.1103/PhysRevLett.96.110404

Klitzing, K. V., Dorda, G., & Pepper, M. (1980). New Method for High-Accuracy Determination of the Fine-Structure Constant Based on Quantized Hall Resistance. *Physical Review Letters*, *45*(6), 494-497. https://doi.org/10.1103/physrevlett.45.494

Konig, M., Wiedmann, S., Brune, C., Roth, A., Buhmann, H., Molenkamp, L. W., Qi, X.-L., & Zhang, S.-C. (2007). Quantum Spin Hall Insulator State in HgTe Quantum Wells. *Science*, *318*(5851), 766-770. https://doi.org/10.1126/science.1148047

Lancaster, T., & Pexton, M. (2015). Reduction and emergence in the fractional quantum Hall state. *Studies In History and Philosophy of Science Part B*, *52*, 343-357. https://doi.org/10.1016/j.shpsb.2015.10.004

Laughlin, R. B. (1983). Anomalous Quantum Hall Effect: An Incompressible Quantum Fluid with Fractionally Charged Excitations. *Phys. Rev. Lett.*, *50*(18), 1395-1398. https://doi.org/10.1103/physrevlett.50.1395

Levin, M., & Wen, X.-G. (2005). Colloquium: Photons and electrons as emergent phenomena. *Reviews of Modern Physics*, *77*(3), 871-879. https://doi.org/10.1103/RevModPhys.77.871

Liu, A. J., Grest, G. S., Marchetti, M. C., Grason, G. M., Robbins, M. O., Fredrickson, G. H., Rubinstein, M., & Olvera de la Cruz, M. (2015). Opportunities in theoretical and computational polymeric materials and soft matter. *Soft Matter*, *11*(12), 2326-2332. https://doi.org/10.1039/c4sm02344g

Liu, C. (1999). Explaining the emergence of cooperative phenomena. *Philosophy of Science* **66**, S92–S106.

Liu, F., Whitsitt, S., Curtis, J. B., Lundgren, R., Titum, P., Yang, Z.-C., Garrison, J. R., & Gorshkov, A. V. (2019a). Circuit Complexity across a Topological Phase Transition. *Phys. Rev. Research 2, 013323 (2020), quant-ph.* https://doi.org/10.1103/PhysRevResearch.2.013323

Liu, P., Williams, J. R., & Cha, J. J. (2019b). Topological nanomaterials. *Nature Reviews Materials, 4*(7), 479-496. https://doi.org/10.1038/s41578-019-0113-4

Maciejko, J., Hughes, T. L., & Zhang, S.-C. (2011). The Quantum Spin Hall Effect. *Annual Review of Condensed Matter Physics, 2*(1), 31-53. https://doi.org/10.1146/annurev-conmatphys-062910-140538

McGinley, M., & Cooper, N. R. (2020). Fragility of time-reversal symmetry protected topological phases. *Nature Physics.* https://doi.org/10.1038/s41567-020-0956-z

McLeish, T. (2019). Soft Matter: An Emergent Interdisciplinary Science of Emergent Entities. In *The Routledge Handbook of Emergence* (pp. 248–264). Routledge. https://doi.org/10.4324/9781315675213-21

McLeish, T., Pexton, M. & Lancaster, T. (2019). Emergence and topological order in classical and quantum systems. *Studies In History and Philosophy of Science Part B* **66**, 155–169.

Moore, J. E. (2010). The birth of topological insulators. *Nature, 464*(7286), 194-198. https://doi.org/10.1038/nature08916

Mourik, V., Zuo, K., Frolov, S. M., Plissard, S. R., Bakkers, E. P. A. M., & Kouwenhoven, L. P. (2012). Signatures of Majorana Fermions in Hybrid Superconductor-Semiconductor Nanowire Devices. *Science, 336*(6084), 1003-1007. https://doi.org/10.1126/science.1222360

Nadj-Perge, S., Drozdov, I. K., Li, J., Chen, H., Jeon, S., Seo, J., MacDonald, A. H., Bernevig, B. A., & Yazdani, A. (2014). Observation of Majorana fermions in ferromagnetic atomic chains on a superconductor. *Science, 346*(6209), 602-607. https://doi.org/10.1126/science.1259327

Nakamura, J., Liang, S., Gardner, G. C., & Manfra, M. J. (2020). Direct observation of anyonic braiding statistics. *Nature Physics, 16*(9), 931-936. https://doi.org/10.1038/s41567-020-1019-1

Nandkishore, R., & Huse, D. A. (2015). Many-Body Localization and Thermalization in Quantum Statistical Mechanics. *Annual Review of Condensed Matter Physics, 6*(1), 15-38. https://doi.org/10.1146/annurev-conmatphys-031214-014726

Nayak, C., Simon, S. H., Stern, A., Freedman, M., & Das Sarma, S. (2008). Non-Abelian anyons and topological quantum computation. *Reviews of Modern Physics, 80*(3), 1083-1159. https://doi.org/10.1103/revmodphys.80.1083

Novoselov, K. S. (2004). Electric Field Effect in Atomically Thin Carbon Films. *Science, 306*(5696), 666-669. https://doi.org/10.1126/science.1102896

Novoselov, K. S., Jiang, Z., Zhang, Y., Morozov, S. V., Stormer, H. L., Zeitler, U., Maan, J. C., Boebinger, G. S., Kim, P., & Geim, A. K. (2007). Room-Temperature Quantum Hall Effect in Graphene. *Science, 315*(5817), 1379-1379. https://doi.org/10.1126/science.1137201

Ovadia, M., Kalok, D., Tamir, I., Mitra, S., Sacépé, B., & Shahar, D. (2015). Evidence for a Finite-Temperature Insulator. *Sci Rep, 5*, 13503. https://doi.org/10.1038/srep13503

Orzel, C., Tuchman, A. K., Fenselau, M. L., Yasuda, M., & Kasevich, M. A. (2001). Squeezed states in a Bose-Einstein condensate. *Science, 291*(5512), 2386-2389. https://doi.org/10.1126/science.1058149

Qi, X.-L., & Zhang, S.-C. (2011). Topological insulators and superconductors. *Reviews of Modern Physics, 83*(4), 1057-1110. https://doi.org/10.1103/revmodphys.83.1057

Sacha, K., & Zakrzewski, J. (2018). Time crystals: a review. *Rep Prog Phys, 81*(1), 016401. https://doi.org/10.1088/1361-6633/aa8b38

Sachdev, S. (2012b). The quantum phases of matter. *arXiv.org, hep-th*. https://arxiv.org/abs/1203.4565v4

Sato, M., & Ando, Y. (2017). Topological superconductors: a review. *Reports on Progress in Physics, 80*(7), 076501. https://doi.org/10.1088/1361-6633/aa6ac7

Savary, L., & Balents, L. (2017). Quantum spin liquids: a review. *Rep Prog Phys, 80*(1), 016502. https://doi.org/10.1088/0034-4885/80/1/016502

Shankar, S., Bowick, M. J. & Marchetti, M. C. (2017). Topological Sound and Flocking on Curved Surfaces. *Physical Review X* 7, 031039.

Silberstein, M., & McGeever, J. (1999). The Search for Ontological Emergence. *The Philosophical Quarterly, 49*(195), 201-214. https://doi.org/10.1111/1467-9213.00136

Song, H., Xiong, C. Z., & Huang, S.-J. (2020). Bosonic crystalline symmetry protected topological phases beyond the group cohomology proposal. *Physical Review B, 101*(16). https://doi.org/10.1103/physrevb.101.165129

Swingle, B., & McGreevy, J. (2016). Renormalization group constructions of topological quantum liquids and beyond. *Physical Review B, 93*(4). https://doi.org/10.1103/physrevb.93.045127

Teller, P. (1986). Relational Holism and Quantum Mechanics. *The British Journal for the Philosophy of Science, 37*(1), 71–81. https://www.jstor.org/stable/686998

Thouless, D. J., Kohmoto, M., Nightingale, M. P., & den Nijs, M. (1982). Quantized Hall Conductance in a Two-Dimensional Periodic Potential. *Physical Review Letters, 49*(6), 405-408. https://doi.org/10.1103/physrevlett.49.405

Tokura, Y., Yasuda, K., & Tsukazaki, A. (2019). Magnetic topological insulators. *Nature Reviews Physics, 1*(2), 126-143. https://doi.org/10.1038/s42254-018-0011-5

Tong, D. (2016). Lectures on the Quantum Hall Effect. *arXiv*. https://arxiv.org/abs/1606.06687v2

Tsui, D. C., Stormer, H. L., & Gossard, A. C. (1982). Two-Dimensional Magnetotransport in the Extreme Quantum Limit. *Physical Review Letters, 48*(22), 1559-1562. https://doi.org/10.1103/physrevlett.48.1559

van Nieuwenburg, E., Liu, Y.-H., & Huber, S. (2017). Learning phase transitions by confusion. *Nature Physics, 13*(5), 435-439. https://doi.org/10.1038/nphys4037

Venderley, J., Khemani, V., & Kim, E. A. (2018). Machine Learning Out-of-Equilibrium Phases of Matter. *Phys Rev Lett, 120*(25), 257204. https://doi.org/10.1103/PhysRevLett.120.257204

Vojta, M. (2003). Quantum phase transitions. *Reports on Progress in Physics, 66*(12), 2069-2110. https://doi.org/10.1088/0034-4885/66/12/r01

Wang, J., & Zhang, S.-C. (2017). Topological states of condensed matter. *Nature Materials, 16*(11), 1062-1067. https://doi.org/10.1038/nmat5012

Wen, X.-G. (1989). Vacuum degeneracy of chiral spin states in compactified space. *Physical Review B, 40*(10), 7387-7390. https://doi.org/10.1103/physrevb.40.7387

Wen, X.-G. (2002). Origin of gauge bosons from strong quantum correlations. *Phys Rev Lett, 88*(1), 011602. https://doi.org/10.1103/PhysRevLett.88.011602

Wen, X.-G. (2003). Quantum order from string-net condensations and the origin of light and massless fermions. *Physical Review D.* https://doi.org/10.1103/PhysRevD.68.065003

Wen, X.-G. (2013). Topological Order: From Long-Range Entangled Quantum Matter to a Unified Origin of Light and Electrons. *ISRN Condensed Matter Physics, 2013*(2), 1-20. https://doi.org/10.1103/PhysRevB.79.214502

Wen, X.-G. (2017). Colloquium: Zoo of quantum-topological phases of matter. *Reviews of Modern Physics, 89*(4), 1064. https://doi.org/10.1103/RevModPhys.89.041004

Wen, X.-G. (2018). Four revolutions in physics and the second quantum revolution—A unification of force and matter by quantum information. *International Journal of Modern Physics B, 32*(26), 1830010. https://doi.org/10.1142/S0217979218300104

Wen, X.-G. (2019). Choreographed entanglement dances: Topological states of quantum matter. *Science, 363*(6429), eaal3099. https://doi.org/10.1126/science.aal3099

Wen, X.-G., & Niu, Q. (1990). Ground-state degeneracy of the fractional quantum Hall states in the presence of a random potential and on high-genus Riemann surfaces. *Phys. Rev., B Condens. Matter, 41*(13), 9377-9396. https://doi.org/10.1103/physrevb.41.9377

Wilczek, F. (2012). Quantum Time Crystals. *Phys Rev Lett, 109*(16), 160401. https://doi.org/10.1103/PhysRevLett.109.160401

Willett, R. L., Shtengel, K., Nayak, C., Pfeiffer, L. N., Chung, Y. J., Peabody, M. L., Baldwin, K. W., West, K. W., &. (2019). Interference measurements of non-Abelian e/4 & Abelian e/2 quasiparticle braiding. *arXiv.* https://arxiv.org/abs/1905.10248v1

Willett, R. L., Nayak, C., Shtengel, K., Pfeiffer, L. N., & West, K. W. (2013). Magnetic-field-tuned Aharonov-Bohm oscillations and evidence for non-Abelian anyons at $\nu = 5/2$. *Phys Rev Lett, 111*(18), 186401. https://doi.org/10.1103/PhysRevLett.111.186401

Yankowitz, M., Chen, S., Polshyn, H., Zhang, Y., Watanabe, K., Taniguchi, T., Graf, D., Young, A. F., & Dean, C. R. (2019). Tuning superconductivity in twisted bilayer graphene. *Science, 363*(6431), 1059–1064. https://doi.org/10.1126/science.aav1910

Zaanen, J., & Beekman, A. J. (2012). The emergence of gauge invariance: The stay-at-home gauge versus local–global duality. *Annals of Physics, 327*(4), 1146-1161. https://doi.org/10.1016/j.aop.2011.11.006

Zeng, B., Chen, X., Zhou, D.-L., & Wen, X.-G. (2019). *Quantum Information Meets Quantum Matter*. Springer New York. https://doi.org/10.1007/978-1-4939-9084-9

Zhang, J., Hess, P. W., Kyprianidis, A., Becker, P., Lee, A., Smith, J., Pagano, G., Potirniche, I. D., Potter, A. C., Vishwanath, A., Yao, N. Y., & Monroe, C. (2017). Observation of a discrete time crystal. *Nature, 543*(7644), 217-220. https://doi.org/10.1038/nature21413

Zhang, S.-C. (2002). To see a world in a grain of sand. *arXiv.org, hep-th*. https://arxiv.org/abs/hep-th/0210162v1

Zhang, T., Jiang, Y., Song, Z., Huang, H., He, Y., Fang, Z., Weng, H., & Fang, C. (2019). Catalogue of topological electronic materials. *Nature, 566*(7745), 475-479. https://doi.org/10.1038/s41586-019-0944-6

3

Emergence of Space

Abstract For a subject that is supposed to unify physics, quantum gravity is notorious for its heated debates. But in the past decade, researchers have found they are often just working different sides of the same jigsaw puzzle. They have united around the idea that space and time are an emergent structure. Black holes are their main case study; the paradox of information loss in these objects is a major focus of research. By securitizing black holes, theorists developed the holographic principle, which, in the specific form known as the AdS/CFT duality, is their leading idea for how space can emerge. The duality has been refined over the years and now gives a central role to quantum entanglement and computational concepts such as complexity and error correction. To be sure, AdS/CFT is hardly the only game in town. Some have sought to extend its framework to more realistic settings; others have pursued entirely different programs such as loop quantum gravity, causal-set theory, and causal dynamical triangulations. The debates may have been transformed, but have certainly not ended.

Keywords General relativity · Quantum gravity · Holographic principle · AdS/CFT duality · Loop quantum gravity · Causal-set theory · Causal dynamical triangulations · Black holes · Black-hole information paradox · Quantum entanglement · Ryu-Takayanagi conjecture · Tensor networks · Quantum error correction · Fuzzball · Circuit complexity · Scrambling · Vasiliev theory

G. Musser, *Emergence in Condensed Matter and Quantum Gravity*,
SpringerBriefs in Physics, https://doi.org/10.1007/978-3-031-09895-6_3

45

3.1   Introduction—Is Spacetime Doomed?

Physicists in the twentieth century were able to describe most of the forces of nature in terms of quantum theories of particles and their interactions. Electrically charged particles such as electrons tug or repel one another by exchanging photons (particles of light), while atomic nuclei are held together by the aptly named gluons. These theories proved so successful that one school of physicists naturally sought to repeat the trick and derive a theory of quantum gravity. We can call them "lumpers": They conceived of gravity as similar to, if more complicated than, the other forces of nature and sought a quantum description in which gravity is mediated by its own particle, the hypothesized graviton. Other physicists could be termed "splitters": They thought gravity is special and pursued alternative approaches to the particulate picture.

As outlined in this chapter, the twenty first-century view of gravity is something of a compromise between these two views. It now seems the lumpers were right that the study of gravity needs to be integrated with the study of matter, and the splitters were right that gravity is a creature unto itself. Some go so far as to say that gravity does not appear on the fundamental menu of nature, but arise at a higher level.

This shift is transforming humanity's conception of space and time. The core insight of Einstein's general theory of relativity—described in more detail in the next section—is that gravity is not a force that propagates through space and time, but a feature of space and time themselves. Earth orbits the sun because the sun warps the space and time around it, guiding our planet's path into a near-circle. Einstein's space and time, then, are not the fixed, preexisting backdrop that Isaac Newton took them to be and that physicists had conventionally assumed since the founding of their discipline.

Adopting an idea that goes back to Newton's contemporary and rival, Gottfried Leibniz, many theorists now conjecture that space emerges when building blocks of some sort assemble themselves into a structure with a certain order or patterning. Time, too, may emerge, but that poses special complications. This chapter will focus on space, which, as we will see, is puzzling enough.

Disagreement between splitters and lumpers, as well as among the various approaches each group has pursued, now centers on what these building blocks are and precisely how they assemble. These approaches differ on whether the underlying physics is a straightforward variant on what we see or something alien. The "atoms" of space might not be spatial, for instance;

they might not exist within space or have size. Looking at them, you might never suspect they could give rise to space.

In fact, spacetime emergence might occur on four distinct levels, according to quantum-gravity researcher Oriti (2018). Level 0 is quantized gravity: the gravitons that theorists traditionally studied. Level 1 introduces new structures—"atoms" of space. Level 2 supposes that the atoms can form structures that are no longer spatiotemporal. Level 3, the highest in Oriti's hierarchy, conjectures that emergence is not just a logical relationship but a dynamical process. Spacetime can undergo phase transitions: It can literally melt or boil.

Making sense of these ideas demands new theoretical tools. The most powerful and popular of these has been "holography," a concept that originated in the study of black holes ('t Hooft, 1994) and has since been applied to the emergence of space and gravity more generally (Bousso, 2002). The bulk of this chapter will concern the development of the program in the 2000s, which suggests that in some sense our 3-D universe may be a projection from a 2-D hologram. The third dimension that is missing from the hologram emerges through the collective behavior of the "atoms" of space. Recent studies that consider holography in terms of network structures (Swingle, 2012a) and redundant encoding of information (Almheiri et al., 2015) have yielded new insights about the building blocks of space and the nature of black holes. The implications of this combined body of research are profound: What is fundamental is the *quantum* and all else—space, gravity, and perhaps time—is derived from that.

Emergent space also suggests new experimental directions. Theorists working on quantum gravity and unification in general have long been criticized for drifting far from the moorings of experiment. These critiques are somewhat unfair. It is not a failing of theory, but a fact of nature, that unification is remote from laboratory conditions. But certainly it is true that theorists acutely feel the need for experimental guidance. If spacetime is emergent, there may be a wealth of new experimental avenues. One of the most exciting—though speculative—possibilities, which relates to the fate of matter falling into black holes, is covered in Sect. 3.3. Quantum-gravity theorists and condensed-matter physicists have also found resonances between models of the emergence of spacetime and the behavior of exotic materials in the lab. Chapter 4 will outline the cross-fertilization between these two seemingly disparate fields, which is yielding both theoretical and experimental fruit for both disciplines.

Another criticism of holography is that its mathematics is most thoroughly developed for a model of a universe that, though similar to our own in many regards, is quite dissimilar in terms of its overall shape. Section 3.4 describes attempts to move beyond these standard formulations of holography to find a model that more closely mirrors our actual cosmos. Most of these ideas work within a particular approach to quantum gravity known as string theory. The final section will summarize several other leading candidate theories for the origin of space and time: loop quantum gravity (Rovelli, 2004), causal set theory (Sorkin, 2009), and causal dynamical triangulations (Loll et al., 2006).

3.1.1 Einstein's Space

The idea that gravity has to do with the nature of space comes from a simple observation: Gravity is universal. Everything falls—in fact, falling is the natural state of affairs. The only reason things ever stop falling is that something gets in the way. Everything falls at the same rate. When you let go of a hammer and a feather, they accelerate toward the floor equally, and the only reason the feather takes longer to hit is air resistance. Gravity, then, dictates the motion of objects in the absence of other forces. It is the backdrop to everything else that happens.

But that is the same function that space has. Space is the backdrop to all events in physics, and the geometric properties of space guide the motion of objects. For instance, we are taught in school that objects following parallel paths on a flat plane never meet. On curved surfaces, these rules get more interesting. Consider the globe. If two airplanes take off from different cities on the equator and begin flying due north along lines of longitude, they will gradually get closer together and, if they make it all the way to the pole, meet. That is the result of Earth's spherical geometry.

The airplane example is closely analogous to what happens when you toss a ball in the air and it drops back to the ground (Unruh, 1997). The ball and the ground follow paths that bring them together. The main difference is that these paths are through 3-D space rather than along a 2-D planetary surface. Similarly, Earth's orbit around the sun is nothing more or less than the result of geometry—this was Einstein's great insight. There is no need to talk about forces acting in space. Space itself is enough.

That doesn't mean it is wrong to think of gravity as a propagating force. To the contrary, the idea of gravitational force flows from the geometric picture under the right circumstances, and it is undeniably useful. But for gaining a deeper understanding of gravity, as well as seeing the full range of phenomena that gravity is capable of, only the geometric picture will do. Scientists routinely develop explanations on multiple levels. All are valid within their respective domains.

Fig. 3.1 Wormholes, also known as Einstein-Rosen bridges, are hypothesized to connect distant parts of spacetime (*Image credit* author.)

What sets the geometric properties, and therefore why Earth orbits the sun rather than speed out into deep space or some other outcome, is the arrangement of matter and energy. Space, like anything else in nature, can act and be acted upon—this was another of Einstein's great insights. The sun, by virtue of its mass, carves a giant valley around it, and the planets glide along the topographic contours. The valley is not directly visible to us (since space itself is invisible), but we can tell it's there by watching how the planets move. Valleys are not the only shapes that space can assume. It can form deep pits or, as Einstein and his colleague Rosen (1935) showed, bridges that link otherwise distant locations—what we now call wormholes (Fig. 3.1).

The operative geometric rules are a bit more involved than those we learn in school, because they are not merely spatial but also temporal. Although this chapter focuses on space, always in the background is that space and time are twin aspects of a unified structure, spacetime. The combination of space and time into a unified structure means that the universe has a fundamental unit of velocity. Physicists call this the speed of light, and it is an inevitable consequence of the fungibility of space and time. You can maintain light speed only if you begin at light speed. That is of little practical use to space travelers accelerating from slower velocities; thus the speed of light is not just a fundamental unit of velocity, but also a fundamental limit to velocity. This limit has crucial consequences for the structure of space, as we shall we.

3.1.2   Hints that Space Is Emergent

As yet physicists have no direct evidence that space is the surface layer of a deeper structure. There is, however, indirect evidence for this line of reasoning that comes from gaps or paradoxes in existing theories. As Einstein was the first to recognize, his general theory of relativity betrays its own limitations. Exhibit A is the black hole—a cosmic sinkhole so dense that nothing that enters its perimeter can ever get back out again. To escape would require accelerating to faster than light, which, as we have seen, is impossible. Gravity overpowers all other forces, compressing the black hole's core to an infinite density, known as a spacetime singularity.

Yet general relativity cannot describe the nature of the singularity. Such breakdowns in other domains of physics signal that a theory must be supplanted by a deeper one (Batterman, 2010). In this case, it is believed that a theory of quantum gravity, uniting quantum theory with general relativity, will be needed to track down to this level. A singularity is also thought to lie at the origin of our universe, and Einstein's theory is similarly unable to say what, if anything, came before the big bang.

3.1.2.1 The Black-Hole Information Paradox

Even if we treat black holes as whole objects, without worrying about what happens at their centers, these cosmic behemoths are puzzling. According to general relativity, they can only grow. Their perimeter or "horizon" is a one-way surface. Matter can enter but never return. For some physicists, that is already troubling. The growth of a black holes is an irreversible process, and nothing is physics is supposed to be truly irreversible.

Things get worse when quantum effects are included in the analysis. The issue came into sharp focus in 1974 when Stephen Hawking found that the horizon destabilizes matter and energy fields, causing black holes to emit radiation and gradually evaporate away (Hawking, 1974). The black hole glows like a burning coal, hinting that, like the coal or any other thermal system, it consists of "atoms" in random motion. That means space itself has an atomic composition, since the black hole is nothing more or less than a highly warped region of space (Padmanabhan, 2015).

Hawking's realization led him to formulate a notorious paradox. Since the outgoing radiation carries no information about the black hole's interior or its contents, it is fully randomized, even more so than the thermal radiation emitted by a hot coal. That means that while highly structured matter can fall into the hole, once the black hole has evaporated away entirely, any information once represented by that matter must disappear, too. That apparently violates a law of both classical and quantum physics that information cannot simply vanish completely. Perhaps it does vanish and our laws will have to be reformulated. But theorists argued in the '80s that the failure of such a basic principle would have conspicuous effects all over the universe, not just in black holes, and no such effects are seen (Banks et al., 1984).

Physicists call this the black-hole information paradox (Preskill, 1992). They and philosophers debate whether it is a true "paradox" in the sense of a direct logical contradiction (Wallace, 2017), but no one doubts that it is a major puzzle. To escape a black hole, information must find a way to evade the fundamental limit on speed (Giddings, 1994). That limit is integral to the structure of spacetime, so spacetime, or at least our conventional view of it, must break down somehow.

3.1.2.2 Quantum Entanglement

Although black holes are the main reason to suspect that space is not all it appears to be, hints come from other areas, too. For instance, gravity lacks a property known as renormalizability—a kind of self-similarity—suggesting that its behavior has to change in a fundamental way on small scales. If so, the structure of spacetime must be transformed at that level. Nongravitational physics provides other clues. If spacetime has no fine-scale structure, then physical processes can occur over any distance, approaching zero. This creates divide-by-zero errors in theories of electromagnetism and other forces.

Furthermore, quantum physics allows for a phenomenon known as entanglement, in which the properties of specially prepared particles remain correlated even though no force or other mechanism passes through the space between them. The effect has captivated and confounded physicists since 1927, when Einstein realized that the correlation was at odds with another feature of quantum physics: randomness (Bacciagaluppi & Valentini, 2009).

The randomness arises because of what physicists call superposition: A quantum particle such as a photon or an electron can be placed into an ambiguous state of not having a specific energy, position, or other property. It is only upon measurement that the particle's properties are forced to snap into a single well-defined value, such as being in one location or possessing a certain value of energy. Before the particle is measured, you cannot predict with certainty where the particle will land; the outcome is random. The equations of quantum theory only allow you to calculate the odds of getting one result or another.

If you have two or more such particles, you can place each into a superposition, so that they will all acquire properties at random. But you can also do something that is unique to quantum physics. You can place the particles into a joint superposition, which links their fates together. The particles will still acquire properties at random, but these properties will correlate. In a standard experiment, you consider a pair of particles, each of which stores a bit of information, 0 or 1. You place them into a joint superposition. If you measure either particle, you will find 0 half the time and 1 half the time. Yet the relation between them is fixed. Depending on the details of how you have set them up, they will be either the same (both 0 or 1) or different (one is 0, the other 1). So they are individually random but collectively organized. Such order amid randomness is impossible in ordinary physics. It violates the basic principle that random outcomes are independent. For instance, if you roll a pair of dice, the two dice may knock into each other, but the sides they land on are unrelated. If the dice always came up doubles, you would rightly suspect that something was awry. And this routinely happens to quantum particles.

Entanglement holds even when the particles have no known connection between them, which can be assured by separating them by vast distances. Either the particles must not be truly random or they must instantly exert an influence over each other, no matter how far apart they may be separated. Neither option is palatable. The latter one appears to defy Einstein's theory of relativity, which implies that no influence can travel faster than light. This has been dubbed the EPR paradox, named for the initials of Einstein, Rosen, and Boris Podosky who wrote a highly influential paper about it in 1935 (Einstein et al., 1935). It is not strictly a paradox, but it certainly is mysterious. No mechanism operating within space can explain these correlations. There are almost as many ideas for what could be happening as physicists thinking about the problem.

Some conjecture that the correlations are rooted in a pregeometric physics that is more fundamental than space. As we shall see in Sect. 2.2, entanglement has been invoked by some physicists searching for a possible recipe for making space from purely quantum ingredients.

The bulk of the remainder of the chapter focuses on research into the emergence of space, but it is worth noting that there are alternative views. For each piece of evidence for the emergence of space from something deeper, there are physicists who suggest it might be accounted for without a wholescale revision to physics. One such program known as "asymptotic safety" hews closely to the traditional notion of the spacetime continuum and adapts methods from the study of other forces, although even it implies a significant change to the microscopic structure of space (Percacci, 2009). But at the moment, emergent space strikes most physicists who work in the area as the likeliest direction.

3.2   Space from Entanglement

3.2.1   The Universe as a Hologram—The AdS/CFT Duality

Perhaps the single most important clue to emergent space is the holographic principle, which has inspired a specific mechanism for how one or more dimensions of space might pop out of deeper physics. The principle was proposed by 't Hooft (1994) and developed in the context of string theory by Leonard Susskind (1995). It builds on the thermodynamic features of black holes touched on in the previous sections when discussing their temperature and ability to emit thermal radiation.

Thermodynamics was founded in the nineteenth century to analyze the behavior of steam engines, and it involves the study of heat, temperature, and entropy—roughly, a measure of the disorder in the arrangement of particles in a system, such as the steam-powered pumps and locomotives of the industrial revolution. The second law of thermodynamics states that the entropy of an isolated system cannot decrease over time; systems tend towards disorder unless an outside influence is acting on them. This principle has proved to be one of the most general in science, extending far beyond its origins in heavy machinery. That black holes can only grow, never shrink, was the first hint that they, too, obey the second law. Black holes may be more complicated to analyze in thermodynamic terms than gas pistons, but the fundamental rules that apply to steam engines also bind them.

When matter falls into a black hole, it must contribute entropy to the cosmic object. Just as the entropy of a gas cylinder is an emergent quantity related to the collective properties of the particles held within, that black holes have entropy is another indication that somehow they, too, are collective phenomena. Bekenstein (1973) argued that the entropy of the black hole grows in proportion to the surface area of its event horizon, rather than—as for most everything else in physics—in proportion to its mass or volume. Thus the number of "atoms" making up a black hole scales with its area, rather than with its mass or its volume. The black hole may look like a three-dimensional system, but it behaves like a two-dimensional one. If you try to store a certain amount of material in a black hole, you find yourself strangely unable to do so.

This situation goes under the name of the holographic principle, because a hologram presents itself to us as a three-dimensional object, but on closer examination is an image produced by a two-dimensional sheet of film. Much the same seems to be true for a black hole. And what goes for black holes goes for the rest of space, too, because black holes are simply an extreme case that proves a general principle.

The best-developed account of how spacetime holography works takes the form of a so-called duality, a relation between two seemingly very different quantities—in this case, linking gravity to quantum processes. Put forward by Maldacena (1999), it provides a specific recipe for how to create space, at least in part. The space concerned is called an anti-de Sitter (AdS) space, named after the early twentieth-century Dutch mathematician, physicist, and astronomer Willem de Sitter, who calculated possible shapes of the universe allowed by Einstein's general relativity. AdS space has several strange properties, such as a fixed outer boundary and a geometry such that parallel lines diverge. It is important to note that, first, the duality remains a conjecture, not yet proved with full mathematical rigor, and second, that we do not appear to live in a universe with the AdS geometry. Rather, our universe is closer to another configuration allowed by Einstein's equations called "de Sitter" space. Nonetheless, as we shall see, the analysis using AdS space still has profound implications.

In one version of Maldacena's duality, which is a fairly direct translation of an ordinary hologram, three-dimensional space is generated by physics in a two-dimensional space (Fig. 3.2). That 2-D space is threaded by fields—souped-up versions of the electromagnetic field—and filled with fairly standard elementary particles. These fields are quantum in nature and described by what is known as a conformal field theory (CFT). Thus the 2-D surface contains relatively familiar (for physicists, at least) quantum laws governing elementary particles. But it does not contain either gravity or strings. When those conformal fields are structured in the right way, however, they generate an additional dimension, which you can think of perpendicular to the original space. This creates a 3-D anti-de Sitter space within the bulk.

What makes the hologram such an apt metaphor is the way it encodes depth. A hologram stores an image as a wave pattern spread out across its surface. There is an inverse relationship between size in the projection and on the film. An individual object, localized in the three-dimensional image, is delocalized on the film; conversely, two widely separated objects overlap on the film. The same inverse relationship holds for spacetime holography.

Not only do you get the emergent dimension, but Maldacena discovered you also get gravity appearing within that dimension, along with strings and black holes. This is where the AdS/CFT duality goes beyond the hologram metaphor. Any process involving particles and fields on the 2-D boundary can be described by an equivalent process involving gravity, black holes and strings in the 3-D bulk, and vice versa. The equations might be exactly the same. But the physical interpretation would be very different.

The AdS/CFT duality also means that tucked inside a theory of matter—quantum field theory—is a theory of gravity and therefore of space. Maldacena did not consciously put gravity into the theory; gravity emerges as a consequence of the properties of that theory. The quantum is automatically gravitational.

Maldacena undertook his analysis within string theory, but since then theorists have shown that the concept is not specifically stringy. It follows from the fundamental symmetries of gravitation as Einstein expressed in his general theory of relativity. Whenever the quantum state of matter within spacetime changes, so must the gravitational field on the boundary of spacetime. This mirroring effect creates an equivalence between the interior and the boundary (Marolf, 2009; Chowdhury et al., 2021).

The duality is technically just an equivalence, allowing physicists a handy way to convert sticky equations involving black holes, strings, and gravity into easier ones involving quantum fields and particles, or vice versa. However, physicists usually interpret it in stronger terms: that the quantum boundary is more fundamental and the interior space is derived from it. Partly this is methodological: they have a better grasp of the nongravitational physics on the boundary than the gravitational physics in the bulk. But it should be said that philosophers have been critical of physicists for taking one side of the duality as more fundamental (Rickles, 2012; Crowther, 2016, pp. 28–29; De Haro et al., 2016).

Better grasp or not, quantum field theory is still very complicated, so extracting gravity and space from it remains a challenge. In developing tools to meet that challenge, theorists kill two birds with one stone: improving theories on both sides of the duality. What gravity theorists learn can be put to use in laboratory studies of materials, and what condensed-matter theorists learn can illuminate the cosmos. This will be discussed further in Chap. 4.

The primary challenge for the past two decades has been to fill in the details of how, precisely, space, and its contents emerges—as physicists put it, to create a "dictionary" between the boundary and the bulk. Through such a dictionary, puzzles such as black holes could be reframed in terms of quantum fields and therefore solved.

The research to fill out the dictionary allowing physicists to mathematically switch from the boundary to the bulk may reveal clues about what is required to give rise to space. Fields on the boundary must be patterned in some way to encode space. A generic state, plucked at random, will not admit

Spacetime Holography

Equations controlling the behaviour of black holes and strings under gravity, in a specific kind of space, are mathematically equivalent to the equations governing the interactions of quantum systems. This relationship suggests that space is in some sense emergent from quantum interactions. This concept is called holography.

a) The black holes and strings live within a hypothetical 3-D anti-de Sitter (AdS) space, governed by gravity.

b) Meanwhile, more familiar particles live on the 2-D boundary of this space, which is threaded by quantum fields (obeying a so-called conformal field theory (CFT)).

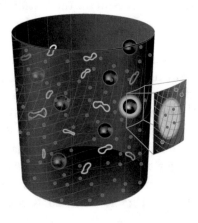

c) The behaviour of specific objects in the 3-D bulk can be calculated by analysing the behaviour of related quantum systems on the boundary, and vice-versa. This is the AdS/CFT duality.

 Particle   String

Quantum field   Black hole

Causal Wedge Paradox

d) Any section on the boundary defines a wedge in the bulk. The behaviour of all objects in the bulk can be said to be controlled by the quantum interactions in the dual section on the boundary.

But some objects in the bulk lie in multiple wedges. This appeared to create a paradox, until it was understood that different sections of the boundary provide complementary and consistent instructions to objects that lie in the overlap region.

JOHN TEMPLETON FOUNDATION FQXi
Created by Maayan Harel for FQXi, the Foundational Questions Institute

of a spatial interpretation (Horowitz & Polchinski, 2009). The next section discusses how this patterning may involve entanglement.

◄**Fig. 3.2** Spacetime holography. Equations controlling the behavior of black holes and strings under gravity, in a specific kind of space, are mathematically equivalent to the equations governing the interaction of quantum systems. This relationship suggests that space is in some sense emergent from quantum interactions. This concept is called holography. **a** The black holes and strings live in a hypothetical 3-D anti-de Sitter (AdS) space, governed by gravity. **b** Meanwhile, more familiar particles live on the 2-D boundary of this space, which is threaded by quantum fields (obeying so-called conformal field theory (CFT)). **c** The behavior of specific objects in the 3-D bulk can be calculated by analyzing the behavior of related quantum systems on the boundary, and vice-versa. This is the AdS/CFT duality. **d** Any section on the boundary defines a wedge in the bulk. The behavior of all objects in the bulk can be said to be controlled by the quantum interactions in the dual section on the boundary. But some objects in the bulk lie in multiple wedges. This appeared to create a paradox, until it was understood that different sections of the boundary provide complementary and mutually consistent instructions to objects that lie in the overlap region (*Image credit* Created by Maayan Harel for the Foundational Questions Institute, FQXi. © FQXi)

3.2.2 Why Entanglement?

For space to emerge, the underlying system must have a special structure to it. The main epiphany of recent years is that this structure is governed by quantum entanglement.

This is an interesting turn of events. As mentioned above, quantum entanglement is one reason that physicists think spacetime might be emergent, and now it turns out that entanglement might explain emergence—it is not only a puzzle, but the answer to a puzzle. Entanglement is a nonspatial effect, so it is a natural substrate for space. At an even simpler level, we expect quantum gravity to be *quantum*, and entanglement is quintessentially quantum.

The key is that entanglement endows quantum systems with holistic qualities, ones that not only cannot be localized, but cannot even be recovered from localized pieces. Because of entanglement, the world around us is full of hidden patterns and structures. By their very nature, they are self-cloaking—in isolation they look random. Only when you look at systems at the collective level do you see their intricate order.

Thus entanglement is a mode of strong emergence—qualitatively new behavior that arises at the collective level. The whole is not the sum of its parts. If the whole were just the sum of its parts, no novelty could arise. Spatiotemporal structure would have to be present in the building blocks. Indeed, a critique of past efforts to describe space as emergent is that they presupposed the very geometry they sought to derive (Meschini et al., 2005). Entanglement may overcome this objection.

The entanglement epiphany happened in slow motion. It has its roots in the 1970s with Bekenstein's and Hawking's work on black holes and entropy, which demonstrated that a geometric quantity (area) can be related to a thermodynamic one (entropy), as described in Sect. 2.1. It so happens there is another concept of entropy, in the context of entanglement, that quantifies the degree of connection between two regions (Jacobson, 1994). Sorkin (1983) argued that the relevant type of entropy when considering black holes is this entropy of entanglement.

Entanglement entropy is related to, though distinct from, thermal entropy. It captures the uncertainty you are left with when you view a system partially. You chop the system in two like a guillotine and then count up the number of particles in one region that are entangled with particles in the other region. Entanglement entropy measures how much information is lost by severing these entangled links if you consider each region in isolation. If you think of the system as a woven cloth, entanglement entropy is the number of threads between the regions. When you cut the cloth, you leave dangling threads, and the number of threads is an indication of how tightly woven the cloth is.

Other theorists besides Sorkin also saw a role for entanglement. In introducing the holographic principle 't Hooft (1994) noted that holography bears a strong resemblance to entanglement. In both phenomena, a system is so highly correlated that, in a sense, the whole is less than the sum of its parts. In the mid-2000s Michael Duff noticed odd mathematical parallels between particle entanglement and black holes in string theory (Duff, 2010).

The field truly exploded in 2006 when Shinsei Ryu and Tadashi Takayanagi conjectured a more granular version of the AdS/CFT duality (Ryu & Takayanagi, 2006; Nishioka et al., 2009). Not only is the boundary equivalent to the bulk, parts of the boundary are equal to parts of the bulk, and entanglement entropy describes this relation.

Consider an area on the boundary. It can be thought of in two ways. First, you can look laterally along the boundary; the matter within the area is entangled with matter elsewhere on the boundary, and the degree of entanglement is measured by the entanglement entropy. Second, you can look down from the area into the interior of the bulk space. The area delineates a wedge of that space (see Fig. 3.2d). The entanglement in the area can be thought of as creating everything that lies inside that wedge.

The wedge is enclosed by a surface with the minimal possible area within the AdS geometry. To visualize this surface, you can imagine blowing a soap bubble anchored on the boundary into the AdS space. The soap bubble is a useful analogy because a bubble seeks to minimize its surface area. By Ryu and Takayanagi's formula, each pair of entangled particles on the boundary corresponds to a unit of area on the AdS slice. This forges a link between entanglement and geometry and, with it, gravity. A given amount of entanglement entropy implies a certain area, which implies a certain spacetime geometry, which implies a certain gravitational field, which implies a certain density of matter and energy (Lin et al., 2015).

The formula has been extended since its original discovery. One version allows variation in time (Hubeny et al., 2007). Another relates entanglement entropy on the boundary not just to area but also to the entanglement entropy of matter in the AdS space, thus providing a unified description of matter and geometry (Faulkner et al., 2013; Engelhardt & Wall, 2015). Yet another rephrases the formula in terms of "threads" that arch from one point on the boundary to another through the bulk (Freedman & Headrick, 2016).

3.2.3 How Entanglement Reproduces Space

In 2009, physicist Mark Van Raamsdonk gave an elegant argument that started a whole subfield investigating how entanglement can knit space together (Van Raamsdonk, 2009). That is to say, if we consider spacetime to be an emergent phenomenon from a more primitive quantum system, then the entanglement within that system is the linchpin.

Consider the AdS/CFT duality. Suppose, Van Raamsdonk argued, the boundary is not entangled. If you have two systems of particles and fields on the boundary, each of these systems will be dual to a space with the AdS geometry. Since the two systems on the boundary are uncorrelated, there is nothing to relate these two spaces. They will be two isolated universes.

But if those systems on the boundary become entangled, then the two spaces cease to be isolated. From the perspective of an observer inhabiting space, the original state corresponds to two disjoint universes, with no way to travel between them. When the systems become entangled, it is as if a wormhole—a spatial tunnel like the ones frequently found in science fiction—opens up between them, and you can move from one to the other. The wormhole has a mouth in each universe, and the interior of the wormhole is shared between the two universes. As the degree of entanglement increases, the wormhole shrinks in length, drawing the universes together until you would not speak of them as two universes anymore. In this way, entanglement accounts for the contiguity of space. Here, Van Raamsdonk was building on earlier work how to describe wormholes within AdS/CFT by Maldacena and Susskind (2003).

The space generated in the bulk has been shown to have other basic required physical features. For instance, within this space, any lump of mass or energy will produce gravity, and that gravity affects all objects. This feature of universality, as we saw, played an important role in Einstein's thinking. Swingle and Van Raamsdonk (2014) have traced it to the universality of entanglement. Other basic concepts such as "points" and "distance" within space might also emerge from entanglement considerations, although it turns out that the more basic the concept, the more subtle the connection (Czech & Lamprou, 2014).

To study how entanglement gives rise to space, in 2009 Swingle imported a powerful mathematical technique from condensed-matter theory known as tensor networks (Swingle, 2012a, b). Tensor networks represent the entanglement structure as a web of relationships among particles or other ingredients (Orús, 2014). They can capture an enormous structure, potentially infinite in extent, at one stoke. Swingle showed that AdS/CFT matches a network with a hierarchy of entanglement analogous to a corporate organizational chart. Climbing or descending the hierarchy has all the qualities of moving within a dimension of space, demonstrating how space might emerge from a system that otherwise lacks it. This line of thinking has made contact with the field of machine learning, since tensor networks are conceptually similar to neural networks. Neural networks seek compact representations of data; likewise, space may be a compact representation of an underlying entanglement structure (Gan & Shu, 2017).

Most impressively, Van Raamsdonk, Swingle, and their colleagues have been able to deduce Einstein's equations of gravitation (Lashkari et al., 2014; Faulkner et al., 2014). Their insight is that entanglement must satisfy certain constraints, and these constraints are just what is needed to create a meaningful geometry and allow objects to tug on one another gravitationally. Jacobson (2016) has performed a similar derivation that interrelates energy, entropy, and geometry. Yasunori Nomura and colleagues (2017) likewise studied how constraints on entanglement entropies endow spacetime with properties that match general relativity.

If these proposals are right, when we feel the tug of gravity, we may really be feeling shifts in the entanglement structure of an underlying system. As residents of spacetime, we are oblivious to the processes that go into creating it. To us, matter and space are separate categories, and gravity is distinct from the other forces of nature. But all have a common origin in the holographic picture. Swingle (2017) has an excellent review.

3.2.4 Space as Error Correction

In 2015 Ahmed Almheiri, Xi Dong, and Daniel Harlow argued the AdS/CFT setup can be thought of as a computer error-correcting code (Almheiri et al., 2015). On the face of it, error correction is a strange analogy. Physicists are not claiming that universe is suffering from errors and must be fixed. Rather, their point is that error correction and emergent space both involve the study of correlations. This has opened a vast toolbox of theoretical ideas and techniques. In fact, error-correction codes have also been taken up in condensed-matter physics to describe quantum phases of matter, as noted in Chap. 2 (Bravyi et al., 2010; Zeng & Zhou, 2014).

Most errors in a computer system, such an electrical glitch, component failure, or ionizing radiation, strike a single location at a time. To protect against loss, computer systems distribute a bit of information across multiple locations in memory. In an ordinary computer, you can make a backup copy and store it off-site, so that a fire or slip of the finger is unlikely to destroy both copies.

In a quantum computer, you have to be more clever. Quantum computers have enhanced processing powers—at least for completing certain tasks—because qubits allow for a richer set of elementary operations. The tradeoff is that quantum information is subject to a wider variety of errors and cannot simply be backed up. You can, however, spread the information out. (An example of this, using a different technique, was described in this chapter when discussing topological quantum computing.) If the components are entangled, the data they store is fully delocalized and resilient against damage.

Reversing the process, you might come across subtle correlations, piece them together, and deduce they represent a single bit. The bits are like jigsaw pieces that have to be found and snapped together. Something like that happens on the boundary in AdS/CFT. There are correlations on all scales. Nearby regions are correlated, as if encoding a bit with a limited degree of redundancy. More distant regions are also correlated, providing extra redundancy. The entire boundary is also correlated, providing the ultimate in security.

Not only does error-correction theory show how bits can be distributed and reassembled, it reveals how they take on the attributes of space. The degree of redundancy can be thought of as a spatial dimension that is not present in the original description of the system. This dimension is envisioned as the axis of a sphere. As the degree of redundancy increases, it is like moving along that dimension, and by the time you reach the maximum redundancy, you have penetrated all the way to the center of space.

The center is a special place in this theory. With anything less than half the boundary, you do not even have partial information about the center—which shows how essential entanglement is to the reconstruction of space. Being equidistant from all points on the boundary, the center of the bulk space is affected by what happens at all those points. If you stood there, your body would, in a deeper sense, be spread out over the entirety of space. As you moved outward, your body would correspond to a smaller portion of the boundary, and by the time you reached the boundary, the boundary and bulk pictures would be equally localized.

3.2.4.1   The Causal Wedge Paradox

This pattern of correlations dispels a paradox related to the bulk and the boundary that was formulated by Alex Hamilton and his colleagues (2006). Imagine a section of the boundary. It defines a wedge of the bulk. Everything inside the wedge is within the range of light signals—or any other causal influence—from the corresponding section of the boundary. For this reason, the wedge is known as a "causal wedge." Just as a city holds sway over a hinterland of suburbs and rural areas, that section of boundary dictates what happens at each point within that wedge. Hamilton and colleagues provided a formula for how to relate quantities in the wedge to a weighted sum of quantities on the corresponding section of the boundary.

They realized that objects in the bulk can lie in more than one wedge. So, the behavior of those observables in the bulk appears to be overdetermined by the boundary. This created a paradox because outright contradictions might arise if quantities in different sections on the boundary gave different instructions to objects lying in the overlap of their respective wedges.

This apparent paradox is resolved by error-correction theory, however, which shows that this redundancy is not a bug, but a feature. It avoids overdetermination and contradiction because different parts of the boundary construct the bulk in different ways, all mutually consistent. It is like taking pictures of a statue from different angles so that you can rebuild it if broken. None of these pictures is enough on its own, but a selection of them gives you enough information, and there is more than one way to make such a selection (Almheiri et al., 2015).

Error-correction theory also expands on the causal-wedge concept. Causal contact is not the only way that that the bulk and boundary remain linked. Entanglement can reach beyond the limits imposed by light travel time. It can shape portions of the bulk space outside the causal wedge, creating a broader "entanglement wedge." (Czech et al., 2012; Headrick et al., 2014; Jafferis et al., 2016; Dong et al. 2016). Entanglement wedges are especially useful for describing black holes. By definition, the interior of a black hole has no causal contact with the outside—light signals cannot escape the hole—yet the interior and the exterior of the black hole can remain entangled (Hartman & Maldacena, 2013; Almheiri, 2018).

3.2.4.2 Modeling Emergence with Qubits

You can make the simplest imaginable model of AdS/CFT using a mere five qubits (Pastawski et al., 2015). An error-correcting code invented in the '90s uses five highly entangled qubits to store one qubit in an error-resistant form. That virtual, or "logical," qubit is smeared out over the five physical qubits. This code has some nice properties. It recovers from any error that strikes one of the five qubits. What is more, it will fill in two lost or corrupted qubits if you know which they are; any three qubits are enough to reconstruct the logical qubit. This is the best error-correction that quantum mechanics allows. It provides an example of correlations that are redundant yet avoid contradictions.

As early as 2000, John Preskill suggested the five-qubit code might be a model of emergent space (Preskill, 2000). The logical qubit can be thought of an atom of space that emerges from the entanglement structure of the physical qubits. If you array the five qubits in a pentagon, you can imagine placing the emergent qubit in the center, giving you a miniature emergent space.

It is really too small to think of geometrically, but does have the properties of the AdS/CFT duality, following the Ryu-Takayanagi formula in a crude sense. Interestingly, the five-qubit code has a high degree of entanglement, but not the highest possible. That is a general principle. For a nice regular space to emerge, the entanglement structure must strike a balance. Only very special states qualify. Most states have too much entanglement and, in them, space is stillborn—it is a giant black hole with no room for anything else (Hayden et al., 2006; Swingle, 2012a).

For all their power, error-correcting codes do run into trouble in that they can capture large-scale aspects of space, but gloss over detailed structure. In fact, this is a general problem with AdS/CFT. The duality does not yet provide a fully developed account of how basic properties of space-time emerge on small scales (Heemskerk et al., 2009). To fix this, physicists have been developing new codes and network models with greater granularity (Yang et al., 2016; Nomura et al., 2018).

3.3   Black Holes

3.3.1   Down the Wormhole—Resolving the Black-Hole Information Paradox

Black holes motivated the study of emergent space, so what do theories of emergent space have to say about them? As discussed earlier in the chapter, the black-hole information paradox, first identified in the 1970s, hinges on the fate of information about objects swallowed by the hole. Hawking realized that when quantum effects are taken into account, black holes slowly emit particles, dubbed Hawking radiation. Eventually, over eons, the black hole will evaporate away entirely, disappearing from the universe and apparently removing all information about its contents in the process. This made most physicists uncomfortable because information should never be able to just vanish from the universe entirely. Information loss would mean that black-hole formation and evaporation is a fundamentally irreversible process.

Physicists attempted to resolve the paradox by suggesting that the information might be carried out of the black hole, encoded in the Hawking radiation. But theory suggested that the radiation is thermal, which means it is random and thus unable to encode any information.

Things are more subtle in the holographic picture, however, offering a potential resolution to the paradox. In the AdS/CFT duality, it is a given that black holes preserve information. That is because black holes emerge from the dynamics of a system of quantum fields, and those fields strictly preserve information. If the fields are in a thermal state—they consist of a gas of particles that has come to equilibrium—they are dual to a black hole. The temperature of the gas is the Hawking temperature of the corresponding black hole.

As described in Sect. 2.3, Van Raamsdonk (2009) noted that AdS/CFT can describe not only black holes, but also wormholes. Maldacena and Susskind (2013) built on Van Raamsdonk's suggestion that a black hole is one mouth of a wormhole connecting two regions of spacetime. They gave their conjecture a name that, to a physicist, is an outrageously clever pun: "ER = EPR." ER stands for an Einstein-Rosen bridge, which is another name for a wormhole, while EPR refers to the aforementioned puzzle of entanglement articulated by Einstein et al. (1935). Like the Ryu-Takayanagi formula, ER = EPR exemplifies how entanglement can create spacetime.

Maldacena and Susskind applied this picture to a new version of the black-hole information paradox that is rephrased in terms of entanglement (Almheiri et al., 2013). The Hawking radiation that is emitted by a black hole is known to be entangled with the interior of the black hole whence it came. Now consider a particle that falls into a black hole. It must eventually return to the outside world in one form or another as the hole evaporates, presumably somehow conveyed by the Hawking radiation that the hole naturally emits. That means the Hawking radiation must be also be entangled with that infalling particle. This creates a problem because there is a principle of "monogamy" in quantum theory that states that an object cannot be fully entangled with two different things at once. The Hawking radiation is already entangled with the interior of the black hole and so cannot also be entangled with the infalling particle without violating this principle. This is the reformulation of the paradox.

But Maldacena and Susskind suggested an audacious escape. They claimed that monogamy would be respected if the Hawking radiation is the interior of the black hole. They appear to be in two different places, but they are really in the same place—the wormhole connects them. In terms of the monogamy metaphor, it's like having an affair with someone only to realize it's your spouse in disguise. What seemed extramarital is really marital. Likewise, what seemed like the outside of a hole is really the inside.

To be sure, the wormhole differs from those in science fiction in one crucial respect: It is not traversable. If you enter one mouth, you are stuck inside. Maldacena and Susskind justify this by noting that entanglement is incapable of conveying a signal from one place to another, so their wormhole should likewise be incapable of channeling matter across space. In later work, they and others did find ways to make the wormhole traversable, but even then it would not provide a short-cut between locations; the path through the wormhole would be longer (Maldacena et al., 2017).

Going still further, Maldacena and Susskind speculated that *all* entangled particles are connected by a wormhole. That would explain how they remain correlated across vast distances. What seems like a vast distance to you is a short distance to them via their private wormhole.

By this logic, any quantum system is dual to some gravitational system. The question is not whether space emerges—it can't help but emerge—but what kind of space emerges. AdS/CFT is special because a fairly conventional geometry (the anti-de Sitter space) emerges rather than a quantum mess that could not support stable structures.

These conjectures have been widely lauded for their profound implications for the emergence of space. It is, however, worth being mindful that they are still speculative; whether they describe the real universe has yet to be determined. One caveat among many is that our universe is not described by an anti-de Sitter space, but as mentioned earlier, is more akin to a de-Sitter space.

Section 4.1 discusses efforts to link these theoretical triumphs pertaining to AdS space to our actual universe.

3.3.2   Island Rule

Although Maldacena and Susskind argued that black holes preserve information, they never really spelled out how. By what means could information escape the deep gravitational pit of a black hole? A string of papers in 2019, culminating in a pair by two teams—an East Coast group of Almheiri, Maldacena, Thomas Hartman, Edgar Shaghoulian, and Amirhossein Tajdini (Almheiri et al., 2019b) and a West Coast group of Geoff Penington, Stephen Shenker, Douglas Stanford, and Zhenbin Yang (Penington et al., 2019)—filled in some of the details and neared a resolution of Hawking's nearly half-century-old paradox. Their analysis also hints at a dramatic example of spacetime emergence.

According to Hawking's original analysis, the radiation emitted by a black hole is entangled with the interior of the hole. As time goes on, the entanglement entropy of the radiation grows, meaning that an increasingly significant amount of information about the system will be lost if the links between the interior and the radiation are cut. The hole sheds mass in the process and eventually evaporates away altogether, at which point the radiation is entangled with—nothing. That violates quantum physics. The radiation presumably carries the imprint of whatever had fallen into the hole, but that imprint is impossible to reconstruct without information from the interior, which is now lost. It is like throwing away half the pieces of a jigsaw puzzle—the puzzle becomes impossible to complete.

For the evaporation process to obey quantum physics, some new effect left out of the analysis must cause the entanglement entropy to stop growing and return to zero (Page, 1993). That way, by the time the hole evaporates away, the radiation is not entangled with the interior any longer; it is a free-standing system. (Note that the thermal entropy still increases throughout, in accordance with the second law of thermodynamics. Only the entanglement entropy must rise and fall.)

The first step in the new analysis is to calculate the entropy of the black hole. The method goes back to a proposal by Rocha (2008) for how to describe black-hole evaporation in the AdS/CFT duality. The usual problem is that a black hole in AdS will not evaporate away completely, but will reach equilibrium with its Hawking radiation. Rocha suggested modifying the conditions on the boundary to absorb radiation and keep the hole out of equilibrium. This fix allowed theorists to follow the evaporation process all the way through (Penington, 2019; Almheiri et al., 2019a). As the hole shrinks, it changes the geometry of the bulk spacetime and, in turn, the entanglement entropy returns to zero. It confirms that no information is ever lost.

The next step is to find where the information ended up. By analyzing the radiation the boundary absorbs, theorists set out to calculate its entropy and see whether it matches that of the hole (Almheiri et al., 2020). Indeed, they confirmed that it does, but in a strange way. It seems that the radiation is never strictly independent of its source, but remains tethered to an "island" within the hole. Although there is no literal tether stretching through space, the two systems are nonetheless linked. There is a duality between them, just as AdS and CFT are alternative descriptions of the same system. From a spatial point of view, the information appears to occupy two different locations.

The system thus defies a spatiotemporal description and so it can be thought of as an example of emergent spacetime. The technique reveals spacetime to be a hall of mirrors where the same object can be both here and there.

If you insist on a spacetime description, you are forced to think in terms of nonlocal effects leaping across the space between the hole and its radiation. In effect, a wormhole connects the two, which seem to be vastly far apart in external space, but are actually adjacent, on account of the wormhole.

In the final step, the East Coast and West Coast teams redid the analysis using basic gravity theory, without assuming the AdS/CFT duality. The result appears to be fully general. If this analysis is correct, the emergence of spacetime does not rely on exotic physics, just on gravitational effects built into Einstein's theory.

The researchers thus claim to have dispelled the information paradox. Black holes do not pose an existential threat to our present theories after all. However, they caution they have not solved all the puzzles of black holes. Their scenario does not specify the precise mechanism of information transfer or identify the black hole's microstates, its "atoms"—leaving room for alternative models that use different approaches to explain the emergence of spacetime, as described in Sect. 3.4.

Even with this proviso, this analysis is technically complicated, involves multiple approximations, and not all theorists accept it. For instance, Hao Geng and colleagues (2021) argue that bleeding off radiation alters the gravitational dynamics and makes the entire setup contrived, so that no broader conclusions can be drawn from it. Others are dubious about the nonlocality that the model entails (Guo et al., 2021; Martinec, 2022).

One competing view, for instance, is that a black hole does not ever truly form. Instead, exotic physics of string theory comes into play and matter collapses into a "string star," or "fuzzball," that is gravitationally stable (Mathur, 2009, 2020). Lacking a horizon, such an object lets information escape freely. Even this model, however, has odd consequences for space—space effectively ceases to exist inside the star. Some breakdown of space seems essential to resolving the paradoxes of black holes, reinforcing the general conclusion that space is an emergent phenomenon.

3.3.3   Complexity and Black Holes

It used to be that black holes were mysterious because of the singularity at their center. Then Susskind directed physicists' attention to the boundaries of black holes. And now he is pointing out yet another puzzling aspect of black holes which his colleagues had missed: its interior volume.

A black hole may look from the outside like a sphere, but it does not have the volume of sphere. In fact, its volume grows over time essentially forever. If the black hole is taken to be one mouth of a wormhole, that wormhole keeps stretching. In the AdS/CFT framework, this is awkward. The black hole is dual to a system of quantum particles and fields. This system reaches equilibrium, whereupon its entropy and other broad-scale properties no longer change. So how can the black hole continue to grow?

Susskind reached into the toolbox of quantum information theory and pulled out a new concept: complexity (Susskind, 2014). Complexity in this context is a computational construct. Susskind invited physicists to think of the dual system as a computer running forever, performing some calculation that becomes ever more intricate. He argued that complexity continues to grow even when entropy plateaus. This can mean, for example, that entanglement spans ever larger regions, until it binds the whole system tightly together—a trend that other measures such as entanglement entropy do not fully capture. Following up, other researchers have worked to prove his conjectures about complexity growth (Haferkamp et al., 2022) and explore what they mean for AdS/CFT (Bouland et al., 2019; Iliesiu et al. 2021).

The black-hole interior may be, in the words of Susskind and colleagues, "the purest form of emergent space" (Brown et al., 2016). Riffing off this theme, they have suggested that gravity in general is a trend toward complexification. It may be a statistical tendency, a product of the behavior of the building blocks of emergent spacetime.

3.3.4 Testing Emergence in the Lab: Black Hole Scrambling

In 2016 Swingle proposed a way to test some of these ideas in the lab, using table-top systems that mimic the behavior of black holes. If the above theories of black holes are right, throwing something into a black hole is like throwing it on a bonfire. It is not strictly impossible to reverse the process, merely very difficult. What makes it tough is that the information embodied by the object is completely dispersed or "scrambled"—scattered over the (still unknown) microscopic constituents of the hole. The information thus becomes delocalized, or "thermalized": so thoroughly mixed into the black hole that it looks like random heat. Ultimately, the information is picked up by the Hawking radiation that the hole emits and ripples into the wider universe by the time the hole has burnt itself to nothingness.

Scrambling is an exotic version of ordinary processes of diffusion or the butterfly effect in chaotic systems. Imagine putting a drop of cream into coffee and watching it disperse. The spreading rate depends on the size of the system, which in turns depends on the number of spatial dimensions. Hayden and Preskill (2007) made the first estimate for how fast a black hole smears out information. They were followed up by Sekino and Susskind (2008) and other teams (Lashkari et al., 2013; Maldacena et al., 2016). The results were startling. The dispersal rate in black holes scales as if the number of dimensions were *infinite*: that is, as if everything is connected directly to everything else, without having to pass through space. This is the maximum possible spreading rate possible in nature and indicates that the black hole is a highly nonlocal object, where the concept of space breaks down.

In terms of duality, scrambling in the black hole in the AdS bulk translates to scrambling in a nongravitational quantum system on the CFT boundary. While black holes cannot be readily (let alone safely) made in the lab, quantum systems are routinely studied in table-top systems and their chaotic dynamics can been analyzed. Such a quantum system could be made up of, say, ions trapped in an electromagnetic cavity, each acting as a qubit that can encode information. This is not intended to be a full model of a black hole, which would require an enormous number of qubits, but a simulation of this particular aspect of their dynamics.

Swingle and colleagues (2016) suggested measuring the scrambling process by running such a system then reversing it, as if pressing rewind, to the starting point. If you tried this trick with air molecules trapped in a room, you'd need to stop them and send them back exactly the way they came.

That would be intractable. But other systems, such as atoms in electromagnetic cavities, can be more tightly controlled. All you need to do to put them into reverse is to switch the sign of an interaction—something experimental physicists know how to do. So, what you do is perturb the system slightly, rewind, and see how close the system is to the original starting point. In a chaotic system, it will in fact be quite far from the starting point—this is the measure of the sensitivity to initial conditions.

Such experiments could measure, for example, how correlations within the system vary with time—do they match the theoretical predictions made for black hole scrambling? If so, this will indicate whether the dynamics of a quantum system are indeed dual to a gravitational system, as conjectured by the AdS/CFT duality. In this sense, they can provide a test of the internal consistency of the theory and check whether theories of emergent spacetime are on the right track.

Swingle and his colleagues suggested several experimental implementations that have since been taken up by a half-dozen labs. Guanyu Zhu and colleagues (2016) proposed a cavity setup. Norman Yao, Gregory Bentsen, and their respective teams have proposed a lattice of cold atoms (Yao et al., 2016; Bentsen et al., 2019).

Results have started to trickle in from groups using ions in a magnetic trap (Gärttner et al., 2016), nuclear magnetic resonance (Li et al., 2016; Wei et al., 2018), a Bose–Einstein condensate on a lattice (Meier et al., 2019), and a quantum computer (Landsman et al., 2019). So far, the systems are still too small to return results that couldn't have been calculated, but they are improving rapidly.

In one especially fascinating application, Adam Brown and his colleagues (2019) have shown how to use rewinding to emulate a traversable wormhole, following earlier theoretical work on the topic by Ping Gao and colleagues (2017). In this set-up, a black hole is simulated by a handful of qubits. Two of these black holes are entangled with each other and represent the two mouths of a wormhole. Now, you mimic sending an independent qubit through the wormhole. First, look at one mouth and feed the traveler qubit into it. You do so by rewinding the mouth system, inserting the traveler qubit—that is, allowing the traveler qubit to interact with the mouth qubits—and fast-forwarding back to the starting point. Then you turn to the second mouth, which is entangled with the first mouth. Fast-forward the system representing this mouth and the traveler qubit should appear there.

Assuming this is successful, the traveler's voyage is easy to understand if you think of the quantum system as dual to a wormhole—the qubit popped out the other wormhole mouth. But it seems downright mysterious without this comparison. It would be as unexpected as stirring creamer into coffee, watching it disperse, and then seeing it recoagulate. If you saw that happen, you would naturally assume that some unseen process is occurring behind the scenes.

At first glance, this experiment resembles quantum teleportation experiments, in which entanglement is used to transfer the quantum information encoded on one qubit in one location to a distant qubit, making it appear as though the qubit has jumped across space (Bouwmeester et al., 1997). Quantum physicists are actually quite used to teleporting qubits in the lab. That in itself is not mysterious (at least not to quantum physicists), and such experiments have little to tell us about the AdS/CFT duality. But those fairly commonplace teleportation tests do not involve the scrambling steps. Rewinding and fast-forwarding the systems should jumble the traveler qubit's information, so that it the traveler will be unrecognizable upon emerging from the second mouth. But if the traveler emerges unscathed, as the proposal predicts, it supports the notion that the AdS/CFT duality holds and that quantum entanglement can be thought of as a wormhole. To be clear, though, it is ultimately a suggestive metaphor rather than a proof of the duality.

Other experiments using benchtop experiments that apply the AdS/CFT conjecture to condensed-matter systems have shed light on the nature of such exotic lab materials. The tests are described in Chap. 4, which describes how physicists are entangling—excuse the pun—ideas from quantum gravity with those from condensed matter to learn more about both disciplines.

3.4   Extending Duality

3.4.1   Going Beyond AdS/CFT

As mathematically powerful as the AdS/CFT duality is, it has a major problem: It is completely unrealistic. Our universe does not have the geometry of AdS space, as mentioned above. On large scales, it is closer to dS space; on small scales, it is nearly unwarped or, as physicists put it, "flat." So, physicists know they need to extend AdS/CFT to other geometries. Recent efforts to dispel the black-hole information paradox, as mentioned above, have done that to a degree. What hinders further progress is that AdS space has special properties—this is why physicists have focused on it. AdS has a boundary: It is like a cosmic snowglobe. And as seen, the equivalence between that boundary and the bulk lies at the heart of all its conclusions.

In contrast, flat and dS space are not bounded by a spatial surface. They do have a boundary of sorts, but it lies in the infinite past and future. In such a setup, the emergent dimension is not space, but time. Changing the size scale on the boundary corresponds to moving within time in the bulk (Strominger, 2001). In a way, this is a step toward the ultimate goal of describing not only space but also time as emergent. But it poses problems that make AdS/CFT a doddle by comparison. Time does not operate on the boundary, so it is hard to formulate a dynamical system of particles and fields there. Without a spatial boundary to nail down the geometry, quantum fluctuations become extreme.

Nonetheless, finding a realistic model of *our* universe requires physicists to overcome these challenges. Eva Silverstein has been working on dS dualities for many years. In 2005 she and colleagues developed a dS/dS duality (Alishahiha et al., 2004). It relates a dS space to a pair of lower-dimensional dS spaces that become entangled. In more recent work with Xi Dong and Gonzalo Torroba, she showed that entanglement entropy matches on both sides of the duality (Dong et al., 2018). As in AdS/CFT, entanglement accounts for spatial contiguity: It takes two disconnected regions and forges a single space out of them. The approach has the significant advantage that the emergent dimension is space, as in AdS/CFT, thereby avoiding the alienness of emergent time.

Another approach is to alter the bulk side of the duality and describe it not by a string theory but by so-called Vasiliev theory, put forward by Fradkin and Vasiliev (1987). It is a highly nonlinear and notoriously complex theory in which matter and geometry are impossible to parse. The proposition is that, early in cosmic history, the nonlinearities lessened, the categories of matter and space become distinct, and it became sensible to think of matter as residing within space. The connection between Vasiliev theory and dS duality was proposed in 2002 (Klebanov & Polyakov, 2002) and has since been developed by Andrew Strominger and his colleagues (Anninos et al., 2011) and by Neiman (2018).

Other approaches make a clean break from AdS/CFT. Carroll and Singh (2018) seek to explain flat space by studying entanglement structure. He and his colleagues begin with a minimalist quantum description of a system, not formulated within spacetime at all. Space emerges entirely from scratch—not just one dimension of space, as in AdS/CFT, but all of them. Not just any system will do: its correlations must be highly redundant—you do not need to specify the full set of correlations, but only a subset from which the rest follow. This allows for a geometric interpretation of correlations. The degree of entanglement defines a notion of spatial distance: The more strongly entangled two regions are, the closer they appear to be (Cao et al., 2017).

Though a very different approach from AdS/CFT duality, it arrives at many of the same conclusions. Entanglement knits spacetime together. The system obeys the holographic principle as a consequence of the requirement of redundancy. Following a strategy similar to Jacobson's derivation of Einstein's equations of gravitation from entanglement constraints, Carroll and his colleagues recovered not just spacetime but the workings of gravity (Cao & Carroll, 2017).

3.4.2 Loop Quantum Gravity

The leading competitor of string theory is known as loop quantum gravity. It subtly but powerfully rethinks the gravitational field. Conventionally, physicists describe gravity in terms of what happens at specific locations. If you drop an object from a certain height, how fast will it accelerate? This seems straightforward, but presents difficulties. What, for example, should be the baseline for measuring height? Loop quantum gravity instead describes gravity in terms of geometric relationships among multiple locations, which are not dependent on arbitrary conventions. Combined with quantum principles, this insight implies that space is made up of quanta, or nuggets, of volume—the "atoms" of space, in this scenario. The quanta form a network, with some quanta linked directly to others. The theory is inspired by a relational view of physics that can be traced to Leibniz. Geometry becomes an exercise in algebra—keeping track of the relations.

These atoms of space might loosely be thought of as the squares of a chessboard, but a naïve chessboard metaphor is problematic because a chessboard is not fully symmetric—it privileges rows and columns over other directions. In loop gravity, networks are placed into quantum superposition, so that the overall system is symmetric. This superposition is the defining process of emergence in loop quantum gravity—what elevates it above thinking of space as granular.

Loop quantum gravity does not accord the holographic principle the central status it has in string theory. It holds that the entropy of a black hole scales with area simply because the perimeter surface mediates our interactions with the hole. Furthermore, the theory suggests that the singularity is, in effect, a subatomic-size star that is capable of storing and releasing information, thereby resolving the information paradox (Ashtekar & Bojowald, 2005; Rovelli & Vidotto, 2014). Loop gravity's structures, so-called spin networks, resemble the tensor networks used to describe patterns of entanglement in AdS/CFT and condensed-matter theory (Singh et al., 2010). Several practitioners have taken inspiration from condensed-matter systems such as the string-net-liquid model described in Chap. 2 (Hamma & Markopoulou, 2011). Loop gravity was also ahead of the curve in borrowing ideas from quantum error correction in its development (Markopoulou, 2006; Smolin, 2006).

Some proponents have argued that loop quantum gravity describes not just gravity but matter as well (Bilson-Thompson et al., 2007). In this view, matter consists of kinks in the fabric of spacetime. The links between the atoms of space are braided, and the ways they can be braided account for the variety of elementary particles. Electric charge, for example, is a twist in a strand. This would give particles their stability—particle braids resist unwinding. It turns out the braiding can be viewed as a quantum error correction code (Vaid, 2019).

Loop gravity has spawned a related but distinct approach known as group field theory. It presumes less prior structure for spacetime than its predecessor theory and allows the putative atoms of space to be created and annihilated. Instead of recovering space in a superposition, group field theory posits other collective effects akin to a vast number of H_2O molecules condensing into liquid water (Gielen et al., 2013). States in this theory are highly entangled, can be expressed as tensor networks, and satisfy the Ryu-Takayanagi formula. These analyses might explain the big bang as a kind of phase transition and allow physicists to follow the genesis of our universe from the preexisting phase (Oriti et al., 2016).

To be sure, the theory has yet to pass a crucial test: that the emergent space derived from the theory matches the one we observe (Crowther, 2016, pp. 194–200). Put simply, loop quantum gravity has emergence, but is it the right kind of emergence?

3.4.3 Causal Sets

Another alternative theory for the emergence of space and time is causal set theory. It starts from a basic insight of general relativity: that spacetime enables a certain set of cause-effect relations and, in turn, you can reconstruct spacetime if you have a complete list of the universe's cause-effect relations among its events. In fact, Einstein argued that cause-effect relations are the operational meaning of spacetime. In causal-set theory, the universe is broken down into a vast number of discrete events that can be placed into a cause-effect sequence. The spacetime continuum is a higher-level description; we assign the events positions in order to satisfy the causal relations (Dowker, 2005; Sorkin, 2009).

Causal set theory attributes black hole entropy and its area scaling to the way that the horizon cleaves the universe in two. In this view, black hole entropy it is not associated with internal states of the hole, but with the horizon itself (Dou, 2003). To achieve area scaling, researchers have found they need to make other assumptions, the physical significance of which remain somewhat mysterious (Sorkin & Yazdi, 2018; Belenchia et al., 2018).

Causal set theory has become the standard-bearer for a fundamentally discrete spacetime. Although the causal network has the right structure, it has a similar problem as loop quantum gravity: It is not clear that it can reproduce observed spacetime (Crowther, 2016, pp. 174–175).

3.4.4 Causal Dynamical Triangulations

A third alternative approach goes by the somewhat convoluted name of causal dynamical triangulations. It builds on an approach to quantum mechanics that Richard Feynman (1948) developed. Known as the path integral, it is the mathematical version of "anything that can happen, does happen." In quantum physics, a particle going from point 'A' to point 'B' takes all possible paths, which are combined in a weighted sum. Similarly, the shape of space-time might be a sum of all possible shapes. Theorists began to apply this concept to gravity in the '50s (Misner, 1957; DeWitt, 1967), and Stephen Hawking championed it in the '70s and '80s (Hawking, 1978). These early efforts ran into various roadblocks. For starters, how do you even define all those possible shapes?

In the late '90s, the practitioners of causal dynamical triangulations got around these difficulties by using a standardized procedure for assembling shapes. Any shape can be created from enough triangles and pyramids, as if building a house from Lego bricks. The process of emergence occurs in the weighted sum as sundry shapes reinforce or offset one another to produce unanticipated outcomes. The spacetime that emerges can have any dimen-sionality and, in fact, can vary in dimensionality (Ambjørn et al., 2005). Spacetime can also go through phase transitions.

Interestingly, conditions must be placed on the assembly process to give rise to a universe like ours, suggesting that emergence requires very specific circumstances (Loll et al., 2006; Loll, 2020). Among these conditions: no wormholes allowed. This marks an important difference between causal dynamical triangulations and some of the approaches mentioned above, which make extensive use of wormholes. For all the progress the technique has made, however, it still struggles to handle situations of practical interest, including black holes (Dittrich & Loll, 2006).

It is worth noting that many rival approaches tackling spacetime emergence, including string theory's AdS/CFT correspondence and loop quantum gravity, have dipped into the same toolbox—in particular, quantum information theory. On the one hand, this consilience might not be surprising, since all these theories are working with the same body of knowledge: quantum theory and general relativity. On the other hand, it might have more profound meaning—for instance, that the various approaches are converging on the same theory from different directions and that the answers to one's problems might be sought in another.

References

't Hooft, G. (1994). Dimensional Reduction in Quantum Gravity. In A. Ali, J. Ellis, & S. Randjbar-Daemi (pp. 284–296). World Scientific. https://arxiv.org/abs/gr-qc/9310026v2

Alishahiha, M., Karch, A., Silverstein, E., & Tong, D. (2004). The dS/dS Correspondence. In AIP Conference Proceedings 743, 393. https://dx.doi.org/https://doi.org/10.1063/1.1848341

Almheiri, A. (2018). Holographic Quantum Error Correction and the Projected Black Hole Interior. *arXiv.org, hep-th.* https://arxiv.org/abs/1810.02055v2

Almheiri, A., Dong, X., & Harlow, D. (2015). Bulk locality and quantum error correction in AdS/CFT. *Journal of High Energy Physics, 04*(4), 163. https://doi.org/10.1007/JHEP04(2015)163

Almheiri, A., Engelhardt, N., Marolf, D., & Maxfield, H. (2019a). The entropy of bulk quantum fields and the entanglement wedge of an evaporating black hole. *Journal of High Energy Physics, 2019a*(12). https://doi.org/10.1007/jhep12(2019a)063

Almheiri, A., Hartman, T., Maldacena, J., Shaghoulian, E., & Tajdini, A. (2019b). Replica Wormholes and the Entropy of Hawking Radiation. *arxiv.org, hep-th.* https://arxiv.org/abs/1911.12333v1

Almheiri, A., Mahajan, R., Maldacena, J., & Zhao, Y. (2020). The Page curve of Hawking radiation from semiclassical geometry. *Journal of High Energy Physics, 2020*(3). https://doi.org/10.1007/jhep03(2020)149

Almheiri, A., Marolf, D., Polchinski, J., & Sully, J. (2013). Black holes: complementarity or firewalls? *Journal of High Energy Physics, 2013*(2), 1-20. https://doi.org/10.1007/JHEP02(2013)062

Ambjørn, J., Jurkiewicz, J., & Loll, R. (2005). The Spectral Dimension of the Universe is Scale Dependent. *Physical Review Letters, 95*(1), 171301. https://doi.org/10.1103/PhysRevLett.95.171301

Anninos, D., Hartman, T., & Strominger, A. (2011). Higher Spin Realization of the dS/CFT Correspondence. *arxiv.org, hep-th.* https://arxiv.org/abs/1108.5735v1

Ashtekar, A., & Bojowald, M. (2005). Black hole evaporation: A paradigm. *arXiv.org, gr-qc.* https://doi.org/10.1088/0264-9381/22/16/014

Bacciagaluppi, G., & Valentini, A. (2009). *Quantum Theory at the Crossroads: Reconsidering the 1927 Solvay Conference.* Cambridge University Press.

Bain, J. (2008). Condensed Matter Physics and the Nature of Spacetime. In (Vol. 4 16, pp. 301–329). Elsevier. https://www.worldcat.org/title/ontology-of-spacetime-ii/oclc/272383153

Banks, T., Susskind, L., & Peskin, M. E. (1984). Difficulties for the evolution of pure states into mixed states. *Nuclear Physics B, 244*(1), 125-134. https://doi.org/10.1016/0550-3213(84)90184-6

Batterman, R. W. (2010). Emergence, Singularities, and Symmetry Breaking. *Foundations of Physics, 41*(6), 1031-1050. https://doi.org/10.1007/s10701-010-9493-4

Bekenstein, J. D. (1973). Black Holes and Entropy. *Physical Review D, 7*(8), 2333-2346. https://doi.org/10.1103/PhysRevD.7.2333

Belenchia, A., Benincasa, D. M. T., Letizia, M., & Liberati, S. (2018). On the entanglement entropy of quantum fields in causal sets. *Classical and Quantum Gravity, 35*(7), 074002. https://doi.org/10.1088/1361-6382/aaae27

Bentsen, G., Hashizume, T., Buyskikh, A. S., Davis, E. J., Daley, A. J., Gubser, S. S., & Schleier-Smith, M. (2019). Treelike Interactions and Fast Scrambling with Cold Atoms. *Physical Review Letters, 123*(13), 2262. https://doi.org/10.1103/PhysRevLett.123.130601

Bilson-Thompson, S. O., Markopoulou, F., & Smolin, L. (2007). Quantum gravity and the standard model. *Class. Quantum Grav., 24*(16), 3975-3993. https://doi.org/10.1088/0264-9381/24/16/002

Bouland, A., Fefferman, B., & Vazirani, U. (2019). Computational pseudorandomness, the wormhole growth paradox, and constraints on the AdS/CFT duality. *quant-ph.* Retrieved from https://arxiv.org/abs/1910.14646v1

Bousso, R. (2002). The holographic principle. *Reviews of Modern Physics, 74*(3), 825-874. https://doi.org/10.1103/RevModPhys.74.825

Bouwmeester, D., Pan, J.-W., Mattle, K., Eibl, M., Weinfurter, H., & Zeilinger, A. (1997). Experimental quantum teleportation. *Nature, 390*(6660), 575-579. https://doi.org/10.1038/37539

Bravyi, S., Hastings, M. B., & Michalakis, S. (2010). Topological quantum order: Stability under local perturbations. *Journal of Mathematical Physics, 51*(9), 093512. https://doi.org/10.1063/1.3490195

Brown, A. R., Roberts, D. A., Susskind, L., Swingle, B., & Zhao, Y. (2016). Holographic Complexity Equals Bulk Action? *Phys. Rev. Lett., 116*(19), 191301. https://doi.org/10.1103/PhysRevLett.116.191301

Brown, A. R., Gharibyan, H., Leichenauer, S., Lin, H. W., Nezami, S., Salton, G., Susskind, L., Swingle, B., & Walter, M. (2019). Quantum Gravity in the Lab:

Teleportation by Size and Traversable Wormholes. *arxiv.org*, *quant-ph*. https://arxiv.org/abs/1911.06314v1

Cao, C., & Carroll, S. M. (2018). Bulk entanglement gravity without a boundary: Towards finding Einstein's equation in Hilbert space. *Physical Review D, 97*(8). https://doi.org/10.1103/physrevd.97.086003

Cao, C., Carroll, S. M., & Michalakis, S. (2017). Space from Hilbert space: Recovering geometry from bulk entanglement. *Physical Review D, 95*(2), 011. https://doi.org/10.1103/PhysRevD.95.024031

Carroll, S. M., & Singh, A. (2018). Mad-Dog Everettianism: Quantum Mechanics at Its Most Minimal. *arXiv.org*, *quant-ph*. https://arxiv.org/abs/1801.08132v1

Chowdhury, C., Godet, V., Papadoulaki, O., & Raju, S. (2021). Holography from the Wheeler-DeWitt equation. hep-th. https://doi.org/10.1007/JHEP03(2022)01

Crowther, K. (2016). *Effective Spacetime*. Springer International Publishing. https://doi.org/10.1007/978-3-319-39508-1

Czech, B., Karczmarek, J. L., Nogueira, F., & Van Raamsdonk, M. (2012). The gravity dual of a density matrix. *Classical and Quantum Gravity, 29*(15), 155009. https://doi.org/10.1088/0264-9381/29/15/155009

Czech, B., & Lamprou, L. (2014). Holographic definition of points and distances. *Phys. Rev. D, 90*(10). https://doi.org/10.1103/physrevd.90.106005

De Haro, S., Mayerson, D. R., & Butterfield, J. N. (2016). Conceptual Aspects of Gauge/Gravity Duality. *Foundations of Physics, 46*(11), 1381-1425. https://doi.org/10.1007/s10701-016-0037-4

DeWitt, B. S. (1967). Quantum Theory of Gravity. II. The Manifestly Covariant Theory. *Physical Review, 162*(5), 1195–1239. https://doi.org/10.1103/PhysRev.162.1195

Dittrich, B., & Loll, R. (2006). Counting a black hole in Lorentzian product triangulations. *Classical and Quantum Gravity, 23*(11), 3849-3878. https://doi.org/10.1088/0264-9381/23/11/012

Dong, X., Harlow, D., & Wall, A. C. (2016). Reconstruction of Bulk Operators within the Entanglement Wedge in Gauge-Gravity Duality. *Phys. Rev. Lett., 117*(2), 021601. https://doi.org/10.1103/PhysRevLett.117.021601

Dong, X., Silverstein, E., & Torroba, G. (2018). De Sitter holography and entanglement entropy. *Journal of High Energy Physics, 2018*(7), 181602. https://doi.org/10.1007/JHEP07(2018)050

Dou, D. (2003). Black-Hole Entropy as Causal Links. *Foundations of Physics, 33*(2), 279-296. https://doi.org/10.1023/a:1023781022519

Dowker, F. (2005). Causal sets and the deep structure of spacetime. In (Vol. 16, pp. 445–464). World Scientific. https://doi.org/10.1142/9789812700988_0016

Duff, M. J. (2010). Black holes and qubits. *CERN Courier, 50*(4), 13–16. https://cerncourier.com/cws/article/cern/42328

Einstein, A., Podolsky, B., & Rosen, N. (1935). Can Quantum-Mechanical Description of Physical Reality Be Considered Complete? *Physical Review, 47*(10), 777-780. https://doi.org/10.1103/PhysRev.47.777

Einstein, A., & Rosen, N. (1935). The Particle Problem in the General Theory of Relativity. *Physical Review, 48*(1), 73-77. https://doi.org/10.1103/PhysRev.48.73

Engelhardt, N., & Wall, A. C. (2015). Quantum extremal surfaces: holographic entanglement entropy beyond the classical regime. *Journal of High Energy Physics, 2015*(1). https://doi.org/10.1007/jhep01(2015)073

Faulkner, T., Guica, M., Hartman, T., Myers, R. C., & Van Raamsdonk, M. (2014). Gravitation from entanglement in holographic CFTs. *Journal of High Energy Physics, 2014*(3). https://doi.org/10.1007/jhep03(2014)051

Faulkner, T., Lewkowycz, A., & Maldacena, J. (2013). Quantum corrections to holographic entanglement entropy. *Journal of High Energy Physics, 2013*(11), 181602. https://doi.org/10.1007/JHEP11(2013)074

Feynman, R. P. (1948). Space-Time Approach to Non-Relativistic Quantum Mechanics. *Reviews of Modern Physics, 20*(2), 367-387. https://doi.org/10.1103/RevModPhys.20.367

Fradkin, E. S., & Vasiliev, M. A. (1987). On the gravitational interaction of massless higher-spin fields. *Physics Letters B, 189*(1-2), 89-95. https://doi.org/10.1016/0370-2693(87)91275-5

Freedman, M., & Headrick, M. (2016). Bit Threads and Holographic Entanglement. *Commun. Math. Phys., 352*(1), 407-438. https://doi.org/10.1007/s00220-016-2796-3

Gan, W.-C., & Shu, F.-W. (2017). Holography as deep learning. *International Journal of Modern Physics D, 26*(12), 1743020. https://doi.org/10.1142/S0218271817430209

Gao, P., Jafferis, D. L., & Wall, A. C. (2017). Traversable wormholes via a double trace deformation. *Journal of High Energy Physics, 2017*(12). https://doi.org/10.1007/jhep12(2017)151

Gärttner, M., Bohnet, J. G., Safavi-Naini, A., Wall, M. L., Bollinger, J. J., & Rey, A. M. (2016). Measuring out-of-time-order correlations and multiple quantum spectra in a trapped ion quantum magnet. *arXiv.org, quant-ph.* https://arxiv.org/abs/1608.08938v2

Geng, H., Karch, A., Perez-Pardavila, C., Raju, S., Randall, L., Riojas, M., & Shashi, S. (2021). Inconsistency of Islands in Theories with Long-Range Gravity. *hep-th.* https://doi.org/10.1007/JHEP01(2022)182

Giddings, S. B. (1994). Quantum Mechanics of Black Holes. *arXiv.org, hep-th.* https://arxiv.org/abs/hep-th/9412138v1

Gielen, S., Oriti, D., & Sindoni, L. (2013). Cosmology from group field theory formalism for quantum gravity. *Phys. Rev. Lett., 111*(3), 031301. https://doi.org/10.1103/PhysRevLett.111.031301

Guo, B., Hughes, M. R. R., Mathur, S. D., & Mehta, M. (2021). Contrasting the fuzzball and wormhole paradigms for black holes. *arXiv.* Retrieved from https://arxiv.org/abs/2111.05295

Haferkamp, J., Faist, P., Kothakonda, N. B. T., Eisert, J., & Yunger Halpern, N. (2022). Linear growth of quantum circuit complexity. *Nat. Phys.* https://doi.org/10.1038/s41567-022-01539-6

Hamilton, A., Kabat, D., Lifschytz, G., & Lowe, D. A. (2006). Holographic representation of local bulk operators. *Phys. Rev. D*, *74*(6). https://doi.org/10.1103/physrevd.74.066009

Hamma, A., & Markopoulou, F. (2011). Background-independent condensed matter models for quantum gravity. *New Journal of Physics*, *13*(9), 095006. https://doi.org/10.1088/1367-2630/13/9/095006

Hartman, T., & Maldacena, J. (2013). Time evolution of entanglement entropy from black hole interiors. *Journal of High Energy Physics*, *2013*(5). https://doi.org/10.1007/jhep05(2013)014

Hartnoll, S. A., & Kovtun, P. K. (2007). Hall conductivity from dyonic black holes. *Physical Review D*, *76*(6). https://doi.org/10.1103/physrevd.76.066001

Hawking, S. W. (1974). Black hole explosions? *Nature*, *248*(5443), 30-31. https://doi.org/10.1038/248030a0

Hawking, S. W. (1978). Spacetime foam. *Nuclear Physics B*. https://www.sciencedirect.com/science/article/pii/0550321378903759

Hayden, P., Leung, D. W., & Winter, A. (2006). Aspects of Generic Entanglement. *Commun. Math. Phys.*, *265*(1), 95-117. https://doi.org/10.1007/s00220-006-1535-6

Hayden, P., & Preskill, J. (2007). Black holes as mirrors: quantum information in random subsystems. *Journal of High Energy Physics*, *09*(0), 120. https://doi.org/10.1088/1126-6708/2007/09/120

Headrick, M., Hubeny, V. E., Lawrence, A., & Rangamani, M. (2014). Causality & holographic entanglement entropy. *Journal of High Energy Physics*, *2014*(12). https://doi.org/10.1007/jhep12(2014)162

Heemskerk, I., Penedones, J., Polchinski, J., & Sully, J. (2009). Holography from conformal field theory. *Journal of High Energy Physics*, *2009*(10), 079-079. https://doi.org/10.1088/1126-6708/2009/10/079

Horowitz, G. T., & Polchinski, J. (2009). Gauge/gravity duality. In *Approaches to Quantum Gravity: Toward a New Understanding of Space, Time and Matter* (ed. D. Oriti), pp. 169–186. Cambridge University Press. https://doi.org/10.1017/cbo9780511575549.012

Hubeny, V. E., Rangamani, M., & Takayanagi, T. (2007). A covariant holographic entanglement entropy proposal. *Journal of High Energy Physics*, *2007*(07), 062-062. https://doi.org/10.1088/1126-6708/2007/07/062

Iliesiu, L. V., Mezei, M., & Sárosi, G. (2021). The volume of the black hole interior at late times. *arXiv*. Retrieved from https://arxiv.org/abs/2107.06286

Jacobson, T. (1994). Black Hole Entropy and Induced Gravity. *arXiv*. https://arxiv.org/abs/gr-qc/9404039v1

Jacobson, T. (2016). Entanglement Equilibrium and the Einstein Equation. *Physical Review Letters*, *116*(20). https://doi.org/10.1103/physrevlett.116.201101

Jafferis, D. L., Lewkowycz, A., Maldacena, J., & Suh, S. J. (2016). Relative entropy equals bulk relative entropy. *Journal of High Energy Physics*, *2016*(6). https://doi.org/10.1007/jhep06(2016)004

Klebanov, I. R., & Polyakov, A. M. (2002). AdS dual of the critical O(*N*) vector model. *Physics Letters B, 550*(3-4), 213-219. https://doi.org/10.1016/s0370-269 3(02)02980-5

Landsman, K. A., Figgatt, C., Schuster, T., Linke, N. M., Yoshida, B., Yao, N. Y., & Monroe, C. (2019). Verified quantum information scrambling. *Nature, 567*(7746), 61-65. https://doi.org/10.1038/s41586-019-0952-6

Lashkari, N., McDermott, M. B., & Van Raamsdonk, M. (2014). Gravitational dynamics from entanglement "thermodynamics". *Journal of High Energy Physics, 2014*(4). https://doi.org/10.1007/jhep04(2014)195

Lashkari, N., Stanford, D., Hastings, M., Osborne, T., & Hayden, P. (2013). Towards the fast scrambling conjecture. *Journal of High Energy Physics, 2013*(4). https://doi.org/10.1007/jhep04(2013)022

Li, J., Fan, R., Wang, H., Ye, B., Zeng, B., Zhai, H., Peng, X., & Du, J. (2016). Measuring out-of-time-order correlators on a nuclear magnetic resonance quantum simulator. *arXiv.org, cond-mat.str-el*. https://arxiv.org/abs/1609. 01246v2

Lin, J., Marcolli, M., Ooguri, H., & Stoica, B. (2015). Locality of Gravitational Systems from Entanglement of Conformal Field Theories. *Phys. Rev. Lett., 114*(22), 221601. https://doi.org/10.1103/PhysRevLett.114.221601

Loll, R. (2020). Quantum gravity from causal dynamical triangulations: a review. *Classical and Quantum Gravity, 37*(1), 013002. https://doi.org/10.1088/1361-6382/ab57c7

Loll, R., Ambjørn, J., & Jurkiewicz, J. (2006). The Universe From Scratch. *Contemporary Physics, 47*(2). https://doi.org/10.1080/00107510600603344

Maldacena, J. (1999). The Large-*N* Limit of Superconformal Field Theories and Supergravity. *International Journal of Theoretical Physics, 38*(4), 1113-1133. https://doi.org/10.1023/a:1026654312961

Maldacena, J. (2003). Eternal black holes in anti-de Sitter. *Journal of High Energy Physics, 2003*(04), 021-021. https://doi.org/10.1088/1126-6708/2003/04/021

Maldacena, J., Shenker, S. H., & Stanford, D. (2016). A bound on chaos. *Journal of High Energy Physics, 2016*(8). https://doi.org/10.1007/jhep08(2016)106

Maldacena, J., Stanford, D., & Yang, Z. (2017). Diving into traversable wormholes. *Fortschr. Phys., 65*(5), 1700034. https://doi.org/10.1002/prop.201700034

Maldacena, J., & Susskind, L. (2013). Cool horizons for entangled black holes. *Fortschritte der Physik, 61*(9), 781-811. https://doi.org/10.1002/prop.201300020

Markopoulou, F. (2006). Towards Gravity from the Quantum. *arXiv.org, hep-th*. https://arxiv.org/abs/hep-th/0604120v1

Marolf, D. (2009). Unitarity and Holography in Gravitational Physics. *Physical Review D, 79*(4), 044010. https://doi.org/10.1103/PhysRevD.79.044010

Martinec, E. J. (2022). Trouble in Paradox. *arXiv*. Retrieved from https://arxiv.org/abs/2203.04947

Mathur, S. D. (2009). The information paradox: a pedagogical introduction. *Classical and Quantum Gravity, 26*(22), 224001. https://doi.org/10.1088/0264-9381/26/22/224001

Mathur, S. D. (2020). The VECRO hypothesis. *International Journal of Modern Physics D, 29*(15), 2030009. https://doi.org/10.1142/s0218271820300098

Meier, E. J., Ang'ong'a, J., An, F. A., & Gadway, B. (2019). Exploring quantum signatures of chaos on a Floquet synthetic lattice. *Phys. Rev. A, 100*(1). https://doi.org/10.1103/physreva.100.013623

Meschini, D., Lehto, M., & Piilonen, J. (2005). Geometry, pregeometry and beyond. *Studies In History and Philosophy of Science Part B, 36*(3), 435-464. https://doi.org/10.1016/j.shpsb.2005.01.002

Misner, C. W. (1957). Feynman Quantization of General Relativity. *Reviews of Modern Physics, 29*(3), 497-509. https://doi.org/10.1103/RevModPhys.29.497

Neiman, Y. (2018). Towards causal patch physics in dS/CFT. *EPJ Web of Conferences, 168*(4), 01007. https://doi.org/10.1051/epjconf/201816801007

Nishioka, T., Ryu, S., & Takayanagi, T. (2009). Holographic Entanglement Entropy: An Overview. *Journal of Physics A: Mathematical and Theoretical, 42*(50), 504008. https://doi.org/10.1088/1751-8113/42/50/504008

Nomura, Y., Rath, P., & Salzetta, N. (2018). Pulling the boundary into the bulk. *Physical Review D, 98*(2), 026010. https://doi.org/10.1103/physrevd.98.026010

Nomura, Y., Salzetta, N., Sanches, F., & Weinberg, S. J. (2017). Toward a holographic theory for general spacetimes. *Physical Review D, 95*(8). https://doi.org/10.1103/physrevd.95.086002

Oriti, D., Sindoni, L., & Wilson-Ewing, E. (2016). Emergent Friedmann dynamics with a quantum bounce from quantum gravity condensates. *Classical and Quantum Gravity, 33*(22), 224001. https://doi.org/10.1088/0264-9381/33/22/224001

Oriti, D. (2018). Levels of spacetime emergence in quantum gravity. *arxiv.org, physics.hist-ph*. Retrieved from https://arxiv.org/abs/1807.04875v1

Orús, R. (2014). Advances on tensor network theory: symmetries, fermions, entanglement, and holography. *The European Physical Journal B, 87*(11), 143. https://doi.org/10.1140/epjb/e2014-50502-9

Padmanabhan, T. (2015). Gravity and is Thermodynamics. *Current Science, 109*(12), 2236–2242. https://doi.org/10.18520/v109/i12/2236-2242

Page, D. N. (1993). Information in black hole radiation. *Physical Review Letters, 71*(23), 3743-3746. https://doi.org/10.1103/PhysRevLett.71.3743

Pastawski, F., Yoshida, B., Harlow, D., & Preskill, J. (2015). Holographic quantum error-correcting codes: toy models for the bulk/boundary correspondence. *Journal of High Energy Physics, 2015*(6), 149. https://doi.org/10.1007/JHEP06(2015)149

Penington, G., Shenker, S. H., Stanford, D., & Yang, Z. (2019). Replica wormholes and the black hole interior. *arXiv.org, hep-th*. https://arxiv.org/abs/1911.11977v1

Penington, G. (2019). Entanglement Wedge Reconstruction and the Information Paradox. *arxiv.org, hep-th*. https://arxiv.org/abs/1905.08255v2

Percacci, R. (2009). Gravity from a Particle Physicist's perspective. *arXiv.org, hep-th*. https://arxiv.org/abs/0910.5167v1

Preskill, J. (1992). Do Black Holes Destroy Information? *arXiv.org*, *hep-th*. https://arxiv.org/abs/hep-th/9209058v1

Preskill, J. (2000). Quantum information and physics: Some future directions. *Journal of Modern Optics*, *47*(2-3), 127-137. https://doi.org/10.1080/09500340008244031

Rickles, D. (2012). AdS/CFT duality and the emergence of spacetime. *Studies In History and Philosophy of Science Part B*. https://doi.org/10.1016/j.shpsb.2012.06.001

Rocha, J. V. (2008). Evaporation of large black holes in AdS: coupling to the evaporon. *Journal of High Energy Physics*, *2008*(08), 075-075. https://doi.org/10.1088/1126-6708/2008/08/075

Rovelli, C. (2004). *Quantum Gravity*. Cambridge University Press.

Rovelli, C., & Vidotto, F. (2014). Planck stars. *International Journal of Modern Physics D*, *23*(12), 1442026. https://doi.org/10.1142/s0218271814420267

Ryu, S., & Takayanagi, T. (2006). Holographic derivation of entanglement entropy from the anti-de Sitter space/conformal field theory correspondence. *Phys. Rev. Lett.*, *96*(18), 181602. https://doi.org/10.1103/PhysRevLett.96.181602

Sekino, Y., & Susskind, L. (2008). Fast scramblers. *Journal of High Energy Physics*, *2008*(10), 065-065. https://doi.org/10.1088/1126-6708/2008/10/065

Singh, S., Pfeifer, R. N. C., & Vidal, G. (2010). Tensor network decompositions in the presence of a global symmetry. *Phys. Rev. A*, *82*(5). https://doi.org/10.1103/physreva.82.050301

Smolin, L. (2006). Generic predictions of quantum theories of gravity. In D. Oriti (Ed.), *Approaches to Quantum Gravity* (pp. 548–570). Cambridge University Press. https://doi.org/10.1017/cbo9780511575549.033

Sorkin, R. D. (1983). On the Entropy of the Vacuum Outside a Horizon. *Tenth International Conference on General Relativity and Gravitation (held Padova, 4–9 July, 1983), Contributed Papers*, 2, 734–736.

Sorkin, R. D. (2009). Light, links and causal sets. *Journal of Physics: Conference Series*, *174*, 012018. https://doi.org/10.1088/1742-6596/174/1/012018

Sorkin, R. D., & Yazdi, Y. K. (2018). Entanglement entropy in causal set theory. *Classical and Quantum Gravity*, *35*(7), 074004. https://doi.org/10.1088/1361-6382/aab06f

Strominger, A. (2001). The dS/CFT correspondence. *Journal of High Energy Physics*, *10*(1), 034. https://doi.org/10.1088/1126-6708/2001/10/034

Susskind, L. (1995). The world as a hologram. *Journal of Mathematical Physics*, *36*(11). https://doi.org/10.1063/1.531249

Susskind, L. (2014). Entanglement Is Not Enough. *arxiv.org*, *hep-th*. Retrieved from https://arxiv.org/abs/1411.0690v1

Swingle, B. (2012a). Entanglement renormalization and holography. *Physical Review D*, *86*(6), 065007. https://doi.org/10.1103/PhysRevD.86.065007

Swingle, B. (2012b). Constructing holographic spacetimes using entanglement renormalization. *arXiv.org*, *1209*, 3304. https://arxiv.org/abs/1209.3304

Swingle, B. (2017). Spacetime from Entanglement. *Annual Review of Condensed Matter Physics*, *9*(1), annurev-conmatphys. https://doi.org/10.1146/annurev-conmatphys-033117-054219

Swingle, B., Bentsen, G., Schleier-Smith, M., & Hayden, P. (2016). Measuring the scrambling of quantum information. *Physical Review A*, *94*(4), 1200. https://doi.org/10.1103/PhysRevA.94.040302

Swingle, B., & Van Raamsdonk, M. (2014). Universality of Gravity from Entanglement. *arXiv.org*, *hep-th*. https://arxiv.org/abs/1405.2933v1

Unruh, W. G. (1997). Time, gravity, and quantum mechanics. In S. F. Savitt(Vol. 1, pp. 23–65). Cambridge University Press. https://doi.org/10.1017/CBO9780511622861.004

Vaid, D. (2019). Quantum Error Correction in Loop Quantum Gravity. *arxiv.org*, *gr-qc*. https://arxiv.org/abs/1912.11725v1

Van Raamsdonk, M. (2009). Comments on quantum gravity and entanglement. *arXiv.org*, *hep-th*. https://arxiv.org/abs/0907.2939v2

Wallace, D. (2017). Why Black Hole Information Loss is Paradoxical. *arXiv.org*, *gr-qc*. https://arxiv.org/abs/1710.03783v1

Wei, K. X., Ramanathan, C., & Cappellaro, P. (2018). Exploring Localization in Nuclear Spin Chains. *Phys. Rev. Lett.*, *120*(7), 070501. https://doi.org/10.1103/PhysRevLett.120.070501

Yang, Z., Hayden, P., & Qi, X.-L. (2016). Bidirectional holographic codes and sub-AdS locality. *Journal of High Energy Physics*, *2016*(1), 253-224. https://doi.org/10.1007/JHEP01(2016)175

Yao, N. Y., Grusdt, F., Swingle, B., Lukin, M. D., Stamper-Kurn, D. M., Moore, J. E., & Demler, E. A. (2016). Interferometric Approach to Probing Fast Scrambling. *arxiv.org*, *quant-ph*. https://arxiv.org/abs/1607.01801v1

Zeng, B., & Zhou, D.-L. (2014). Topological and Error-Correcting Properties for Symmetry-Protected Topological Order. *arxiv.org*, *quant-ph*. https://arxiv.org/abs/1407.3413v1

Zhu, G., Hafezi, M., & Grover, T. (2016). Measurement of many-body chaos using a quantum clock. *Phys. Rev. A*, *94*(6). https://doi.org/10.1103/physreva.94.062329

4

Lateral Thinking—The Holographic Principle in Condensed Matter

Abstract A two-way flow of ideas has been a boon to physicists in recent years. Condensed-matter theorists have offered their input into the puzzles of quantum gravity, and gravity researchers have returned the favor by introducing them to the holographic principle. This principle shows the equivalence of very different mathematical formulations; problems that are hard in one formulation can be easier in the other. That makes it an important tool for understanding material systems at hot and cold extremes of the temperature scale. And it further demonstrates the universality of emergence concepts.

Keywords Holographic principle · AdS/CFT duality · SYK model · Quark-gluon plasma · Strange metals

4.1 Introduction

Explaining metals and ceramics by comparing them to a black hole sounds more like a comedy punchline than a serious research proposal. But the complexity of such seemingly ordinary materials is so great that physicists have started to realize that a black hole may be easier to understand by comparison. As described in Chap. 2, these materials can restructure themselves into quantum, topological, and nonequilibrium states that traditional theories don't know what to do with. During these exotic phase transitions,

© The Author(s), under exclusive license to Springer Nature
Switzerland AG 2022
G. Musser, *Emergence in Condensed Matter and Quantum Gravity,*
SpringerBriefs in Physics. https://doi.org/10.1007/978-3-031-09895-6_4

vast numbers of particles synchronize their behavior. Even more perplexingly, the very notion of a particle can break down, and quasiparticles with fractional charges, along with other collective structures, can emerge.

In the 1990s gravity theorists studying black holes came up with the holographic principle and, with it, the concept of duality, as discussed in Chap. 3. The Anti-deSitter/Conformal Field Theory (AdS/CFT) duality forged a connection between the workings of spacetime and of ordinary matter (see Fig. 3.2). On one side of the duality stands a universe that is idealized but not altogether unlike ours: a vast volume of space governed by the force of gravity. On the other side is a complex material system, with particles and fields obeying quantum rules. It does not experience the force of gravity and can be thought of a vast shell encasing the first universe.

What the material system lacks in spatial depth, it makes up for in dense complexity, so that the two sides are equally complicated. A spatial relation in the bulk volume translates into a quantum relation on the surface—mathematically, they are identical. Whatever exists in one has an exact counterpart in the other. Because the two sides are dissimilar yet equivalent, the relation between them is a powerful theoretical tool. Complex problems on one side can become mathematically simple to solve on the other. For instance, physicists may struggle to understanding the equations describing a material system filled with heaving masses of particles, but according to the AdS/CFT duality the space to which it is equivalent may be a perfect vacuum that is far easier to handle theoretically. Highly quantum behavior on one side becomes a straightforward classical calculation on the other. Symmetries and behaviors that are hard to see in one become obvious in the other.

Chapter 3 delved into how gravity theorists apply the holographic principle to describe the emergence of spacetime from a quantum system. But in recent years, condensed-matter theorists have turned the logic on its head. They can use the duality to translate a real-world laboratory system, with complicated quantum interactions, to a simpler spatial system—specifically, to a hypothetical black hole in the AdS/CFT bulk volume. (This theoretical black hole might look absolutely nothing like the real cosmic objects studied by astronomers. Gravity is negligible in these idealized models and there is no risk of being sucked to your death. What matters is that the equations are easier to calculate.)

It certainly isn't a comparison that would have occurred to matter physicists if it hadn't been for their gravity colleagues. But it turns out to be a natural metaphor because black holes have a well-defined temperature and can undergo changes as their temperature increases—changes that can be associated with changes of state. This insight has given physicists a new route to analyzing plasmas of nuclear particles as well as quantum phases, including strange metals and high-temperature superconductors, as described in Sect. 4.2. Crucially, this metaphor provides mathematical tools that work even if the AdS/CFT is not true in the universe at large. Even the fiercest critic of string theory can get on board with applying its techniques to material systems. In a further demonstration of the interdisciplinary flow of ideas, condensed-matter physicists have flipped the duality around, as described in Sect. 4.3, to develop the idealized "SYK model" that uses a description of matter to explain how space may emerge.

4.2 Dualing Theories

4.2.1 Nuclear Plasmas and Holography

One of the earliest examples that suggested that the holographic principle could be useful in real-world contexts came in 2005. The Relativistic Heavy Ion Collider (RHIC) is a particle accelerator, located at the Brookhaven Laboratory in New York, that creates plasmas at a temperature of 2 trillion degrees Celsius or higher—hot enough not only to dissolve atoms, but to break protons and neutrons into their constituent quarks and gluons, creating a state of matter that existed in the early universe. These plasmas display several mysterious features. Not only do the fireballs behave more like a liquid than a gas, their viscosity is strangely low and independent of composition (Fig. 4.1).

It turns out that thinking of these plasmas as a kind of black hole can resolve some of these puzzles. That insight came about somewhat serendipitously when two former undergraduate dorm-mates at Moscow State University, Dam Thanh Son and Andrei Starinets, met up for a reunion in New York, where they had both found research positions. Son was now working on plasmas, while Starinets specialized in string theory—the framework described in Chap. 3 which posits that elementary particles are made up of tiny vibrating threads of energy—and had begun working on the AdS/CFT duality. They noticed an uncanny resemblance between the equations they were using to describe the totally disparate physical systems they were studying (Merali, 2011). With colleagues, they used the duality to translate the viscosity of a plasma into a process involving the scattering of gravitational waves off a black hole (Kovtun et al., 2005) and to predict the plasma's shear viscosity (Policastro et al., 2001). Their prediction was confirmed by measurements made at RHIC in 2008 (Luzum & Romatschke, 2008). The theory provides a tidy explanation of why viscosity does not vary

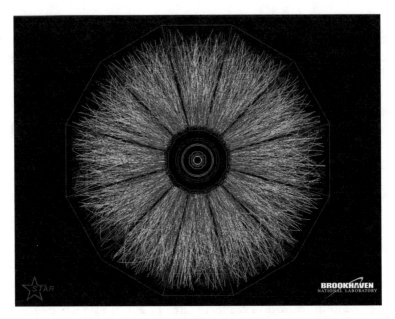

Fig. 4.1 A quark-gluon plasma briefly appears when gold nuclei are collided in the STAR experiment at RHIC, creating a fireball. Its properties are reconstructed from particle tracks (*Image credit* Brookhaven National Laboratory. Shared under Creative Commons license CC BY 2.0.) [https://commons.wikimedia.org/wiki/File:View_of_gold_ions_collision.jpg]

with composition: Black holes look the same no matter what collapsed to form them, so, thanks to the duality, the same must be true of any system to which they are equivalent.

This success notwithstanding, progress has stalled on using duality to understand these fireballs further. The technique really works only for idealized systems, not necessarily the plasmas seen at RHIC or those created at the more energetic Large Hadron Collider, near Geneva, straddling the border between France and Switzerland (DeWolfe et al., 2014; Erlich, 2015). But similar ideas are having unequivocal success at the other end of the temperature spectrum.

4.2.2   Quantum Phases and Holography

In 2007 Subir Sachdev and his collaborators began to apply the AdS/CFT duality to the frigid systems that can metamorphose into quantum phases of matter (Herzog et al., 2007). They and others have developed theories of various condensed-matter phenomena described in Chap. 2, with differing degrees of success, including the Hall effect (Hartnoll & Kovtun,

2007), superconductivity (Hartnoll et al., 2008), the quantum Hall effect (Bayntun et al., 2011), the superfluid-insulator phase transition (Witczak-Krempa et al., 2014) and Mott insulators (Andrade et al., 2018). In addition, a thermoelectric phenomenon called the Nerst effect can also be modeled in this way (Hartnoll et al., 2007).

While the mechanisms behind those effects are already reasonably well understood, translating these material systems into equivalent spatial systems can provide a more intuitive way of understanding their odd behavior, as well as help to solve some of their outstanding puzzles. The superfluid-insulator phase transition is an especially interesting example (Lunts et al., 2015). It has an elegant geometric description when translated to the holographic space. The insulator corresponds to complete fragmentation of the space, as if no part communicated with any other. The superfluid, by contrast, corresponds to a holistic or nonlocal connection of the entire space; in the material, this presents as a high degree of entanglement. The critical point, at which the insulator changes to a superfluid, corresponds to splitting the difference: spatial subsystems can communicate, but only through local processes.

Sachdev (2010b, 2015) has also taken steps to apply holography to strange metals. The defining property of strange metals is that their electrical resistance at very low temperatures is directly proportional to temperature. A holographic analysis reveals that this unexpected arises because these materials are fluid-like. Their electrical resistance arises because of their viscosity (Davison et al., 2014; Zaanen, 2019). Resistance and viscosity are examples of what physicists call transport properties, which describe the movement of charge, energy, and other basic quantities. They are quintessential examples of emergence, behaving in simple ways in spite of the messiness of the material system. Understanding why is a major outstanding problem in condensed-matter theory (White, 2021).

Duality also makes short work of a difficulty that affects electrons and other particles in its class, known as fermions. Probabilities for their behavior can be negative as well as positive, and in many cases the two negate each other. But they will cease to negate if one of the probabilities changes ever so slightly. An impossible process might become possible, or even probable. This extreme sensitivity makes these particles very difficult to approximate—an instability is known as the "sign problem." But the problem simplifies if, by using a duality, you translate the fermion behavior into a problem in classical physics (Čubrović et al., 2009).

Perhaps the greatest prize, as noted in Chap. 2, would be to understand the mechanism behind high-temperature superconductivity. Herzog (2009) has taken steps toward that by using holography to describe superconductors as a transition from one type of black hole to another. Sachdev and others have written cogent reviews of how dualities are helping with condensed matter (McGreevy, 2010; Sachdev, 2012; Hartnoll et al., 2016).

4.3   The SYK Model—Creating Spacetime from Particles

Since holography has proven useful at illuminating weird condensed-matter phenomena, it seems reasonable to see what may be gleaned by turning the duality around and shining a light in the opposite direction. What can physicists hoping to understand the origins of space and time learn from condensed-matter systems?

In 1992 Sachdev and his then student Jinwu Ye proposed a model to explain magnets and strange metals—the odd properties of which, as first discussed in Chap. 2, suggested a hitherto unsuspected quantum–mechanical ordering (Sachdev & Ye, 1993). At around the same time, another pair of theorists, Tevian Dray and Gerard 't Hooft, were thinking about a completely different problem: how to escape a black hole (Dray & 't Hooft, 1985). Shock waves, they proposed, might open a door to solving the black-hole information paradox. This puzzle (laid out in detail in Sect. 3.1.3) involves the fate of information about the objects swallowed by a black hole when the black hole evaporates away to nothing. Many physicists thought that this information may leak out in the form of so-called Hawking radiation emanating from the perimeter of the hole. But according to the usual analysis, this radiation is thermal, meaning that it is randomized and cannot directly encode information.

Dray and 't Hooft's idea was that shock waves are set off when objects fall into a black hole and these shocks could cause outgoing Hawking radiation to be slightly delayed. That would allow information to make a short hop across the horizon and be carried off into the universe by the Hawking radiation.

These two lines of thinking, involving strange metals on the one hand and black holes on the other, seemed to have nothing whatsoever to do with each other. But, as described in the previous section, by 2010 Sachdev had applied the holographic principle to his model of strange metals, showing that they are equivalent to black holes, and using the duality to help calculate their electrical resistance (Sachdev, 2010a, b). Sachdev and Ye then realized that the duality could be turned around to help ease the black-hole calculations. They developed a model that reproduced the behavior of matter near a black hole. Soon after, Alexei Kitaev modified the model—which became known as the Sachdev-Ye-Kitaev or "SYK" model—and applied it to tracking objects falling into a black hole. He found that the shock waves predicted by Dray and 't Hooft appear in this model (Kitaev & Suh, 2018).

The ideas burst onto the quantum-gravity scene in a series of talks that Kitaev gave in spring 2015. He began disarmingly by describing how he had come up with his model a few days earlier and finished his calculations just a few hours before the talk (Kitaev, 2015a, b). Kitaev envisioned a vast number of particles, each interacting with all the others. The interactions do not depend on space; in effect, the particles are all piled on a single point in space. The interactions are also randomized. Four particles can interact at a time in a random way, and the overall effect is the sum total of such interactions. Through the holographic duality, this system is dual to a gravitational system. So, as with the examples involving the AdS/CFT conjecture given in Chap. 3, it can be used to explain how gravity and space might arise from quantum systems. But unlike most other models, the SYK model is fully solvable mathematically, so it quickly became a workhorse for studying spacetime emergence.

The model has many other nice properties. It involves a simpler quantum system than other instances of the holographic principle. It consists of discrete particles rather than a continuum of fields (Bonzom et al., 2017). Gravity, in this model, behaves according to an augmented version of Einstein's general theory of relativity developed by Jackiw (1985) and Teitelboim (1983). The newly branded SYK model was immediately taken up by leading string theorists (Maldacena & Stanford, 2016; Polchinski & Rosenhaus, 2016; Witten, 2016).

A specific use of the SYK model has been to rethink the black-hole information paradox. As discussed in Sect. 3.3.1, physicists made tremendous progress in 2019 to dissolve the paradox, and the SYK model helped by providing an initial theoretical check on their results (Penington, 2019; Penington et al., 2019). It also has allowed them to move beyond the question of information to other puzzles of black holes, such as their energy. Using the SYK model, Phil Saad, Stephen Shenker, Douglas Stanford, and their colleagues calculated that the energy of the black hole can take on a ladder of values—a ladder, that is, whose rungs are not evenly spaced (Cotler et al., 2017). The researchers ran their simulation multiple times with random variations, and certain aspects of the spacing remain constant, while others fluctuated wildly. The meaning of this is not yet quite clear, but the authors think the simulation may be tracking the emergence of a geometry that contains wormholes—hypothetical tunnels from one place to another—from one that lacks them (Saad et al., 2018).

The SYK model is tractable not only mathematically but also experimentally. It could be implemented on a quantum computer (Babbush et al., 2019), in a system of cold atoms corralled by light waves (Danshita et al., 2017), and in magnetic vortices in a superconductor (Pikulin & Franz, 2017). although all these experiments are somewhat beyond the current state of the art.

In the end, these mathematical associations might hint at a deep underlying commonality between matter and space. Or they might simply be a useful metaphor. For the time being, it doesn't really matter. At the very least, gravity and matter theorists have used these mathematical associations to get out of mental ruts. But potentially holography may eventually enable physicists to investigate fundamental aspects of spacetime in a controlled way using condensed-matter systems in the laboratory. And holographic techniques may one day reveal states of matter that condensed-matter physicists have never dreamt of (Sachdev, 2012).

References

Andrade, T., Krikun, A., Schalm, K., & Zaanen, J. (2018). Doping the holographic Mott insulator. *Nature Phys*, *14*(10), 1049-1055. https://doi.org/10.1038/s41 567-018-0217-6

Babbush, R., Berry, D. W., & Neven, H. (2019). Quantum simulation of the Sachdev-Ye-Kitaev model by asymmetric qubitization. *Phys. Rev. A*, *99*(4). https://doi.org/10.1103/physreva.99.040301

Bayntun, A., Burgess, C. P., Dolan, B. P., & Lee, S.-S. (2011). AdS/QHE: towards a holographic description of quantum Hall experiments. *New Journal of Physics*, *13*(3), 035012. https://doi.org/10.1088/1367-2630/13/3/035012

Bonzom, V., Lionni, L., & Tanasa, A. (2017). Diagrammatics of a colored SYK model and of an SYK-like tensor model, leading and next-to-leading orders. *Journal of Mathematical Physics*, *58*(5), 052301. https://doi.org/10.1063/1.498 3562

Cotler, J. S., Gur-Ari, G., Hanada, M., Polchinski, J., Saad, P., Shenker, S. H., Stanford, D., Streicher, A., & Tezuka, M. (2017). Black holes and random matrices. *Journal of High Energy Physics*, *2017*(5). https://doi.org/10.1007/jhe p05(2017)118

Čubrović, M., Zaanen, J. & Schalm, K. String Theory, Quantum Phase Transitions, and the Emergent Fermi Liquid. *Science* **325**, 439-444 (2009).

Danshita, I., Hanada, M., & Tezuka, M. (2017). Creating and probing the Sachdev–Ye–Kitaev model with ultracold gases: Towards experimental studies of quantum gravity. *2017*(8). https://doi.org/10.1093/ptep/ptx108

Davison, R. A., Schalm, K., & Zaanen, J. (2014). Holographic duality and the resistivity of strange metals. *Physical Review B, 89*(24), 231. https://doi.org/10.1103/PhysRevB.89.245116

DeWolfe, O., Gubser, S. S., Rosen, C., & Teaney, D. (2014). Heavy ions and string theory. *Progress in Particle and Nuclear Physics, 75*, 86-132. https://doi.org/10.1016/j.ppnp.2013.11.001

Dray, T., & 't Hooft, G. (1985). The gravitational shock wave of a massless particle. *Nuclear Physics B, 253*, 173-188. https://doi.org/10.1016/0550-3213(85)90525-5

Erlich, J. (2015). An introduction to holographic QCD for nonspecialists. *Contemporary Physics, 56*(2), 159-171. https://doi.org/10.1080/00107514.2014.942079

Hartnoll, S. A., & Kovtun, P. K. (2007). Hall conductivity from dyonic black holes. *Physical Review D, 76*(6). https://doi.org/10.1103/physrevd.76.066001

Hartnoll, S. A., Kovtun, P. K., Müller, M., & Sachdev, S. (2007). Theory of the Nernst effect near quantum phase transitions in condensed matter and in dyonic black holes. *Phys. Rev. B, 76*(14). https://doi.org/10.1103/physrevb.76.144502

Hartnoll, S. A., Herzog, C. P., & Horowitz, G. T. (2008). Building a holographic superconductor. *Phys. Rev. Lett., 101*(3), 031601. https://doi.org/10.1103/PhysRevLett.101.031601

Hartnoll, S. A., Lucas, A., & Sachdev, S. (2016). Holographic quantum matter. *arXiv.org*. https://arxiv.org/abs/1612.07324v3

Herzog, C. P., Kovtun, P., Sachdev, S., & Son, D. T. (2007). Quantum critical transport, duality, and M theory. *Phys. Rev. D, 75*(8). https://doi.org/10.1103/physrevd.75.085020

Herzog, C. P. (2009). Lectures on holographic superfluidity and superconductivity. *Journal of Physics A: Mathematical and Theoretical, 42*(3), 343001. https://doi.org/10.1088/1751-8113/42/34/343001

Jackiw, R. (1985). Lower dimensional gravity. *Nuclear Physics B, 252*, 343-356. https://doi.org/10.1016/0550-3213(85)90448-1

Kitaev, A. (2015a). A simple model of quantum holography (part 1) Kavli Institute for Theoretical Physics Santa Barbara. Retrieved 2020-02-13 from https://online.kitp.ucsb.edu/online/entangled15/kitaev/

Kitaev, A. (2015b). Hidden correlations in the Hawking radiation and thermal noise. Santa Barbara. Retrieved 2020-02-14 from https://online.kitp.ucsb.edu/online/joint98/kitaev/

Kitaev, A., & Suh, S. J. (2018). The soft mode in the Sachdev-Ye-Kitaev model and its gravity dual. *Journal of High Energy Physics, 2018*(5). https://doi.org/10.1007/jhep05(2018)183

Kovtun, P., Son, D. T., & Starinets, A. (2005). Viscosity in Strongly Interacting Quantum Field Theories from Black Hole Physics. *Physical Review Letters, 94*(11), 111601. https://doi.org/10.1103/PhysRevLett.94.111601

Lunts, P., Bhattacharjee, S., Miller, J., Schnetter, E., Kim, Y. B., & Lee, S.-S. (2015). Ab initio holography. *Journal of High Energy Physics, 2015*(8), 253. https://doi.org/10.1007/JHEP08(2015)107

Luzum, M., & Romatschke, P. (2008). Conformal relativistic viscous hydrodynamics: Applications to RHIC results atsNN=200GeV. *Physical Review C, 78*(3).
https://doi.org/10.1103/physrevc.78.034915

Maldacena, J., & Stanford, D. (2016). Remarks on the Sachdev-Ye-Kitaev model.
Phys. Rev. D, 94(10). https://doi.org/10.1103/physrevd.94.106002

McGreevy, J. (2010). Holographic Duality with a View Toward Many-Body Physics.
Advances in High Energy Physics, 2010(3-4), 1-54. https://doi.org/10.1155/2010/
723105

Merali, Z. (2011). Collaborative physics: String theory finds a bench mate. *Nature,
478*(7369), 302-304. https://doi.org/10.1038/478302a

Penington, G., Shenker, S. H., Stanford, D., & Yang, Z. (2019). Replica wormholes
and the black hole interior. *arXiv.org, hep-th.* https://arxiv.org/abs/1911.11977v1

Penington, G. (2019). Entanglement Wedge Reconstruction and the Information
Paradox. *arxiv.org, hep-th.* https://arxiv.org/abs/1905.08255v2

Pikulin, D. I., & Franz, M. (2017). Black Hole on a Chip: Proposal for a Physical
Realization of the Sachdev-Ye-Kitaev model in a Solid-State System. *Phys. Rev.
X, 7*(3). https://doi.org/10.1103/physrevx.7.031006

Polchinski, J., & Rosenhaus, V. (2016). The spectrum in the Sachdev-Ye-Kitaev
model. *Journal of High Energy Physics, 2016*(4), 1-25. https://doi.org/10.1007/
jhep04(2016)001

Policastro, G., Son, D. T., & Starinets, A. O. (2001). Shear Viscosity of Strongly
Coupled N=4 Supersymmetric Yang-Mills Plasma. *Physical Review Letters, 87*(8).
https://doi.org/10.1103/PhysRevLett.87.081601

Saad, P., Shenker, S. H., & Stanford, D. (2018). A semiclassical ramp in SYK and
in gravity. *arXiv.* https://arxiv.org/abs/1806.06840v2

Sachdev, S., & Ye, J. (1993). Gapless spin-fluid ground state in a random quantum
Heisenberg magnet. *Phys. Rev. Lett., 70*(21), 3339-3342. https://doi.org/10.
1103/PhysRevLett.70.3339

Sachdev, S. (2010a). Holographic metals and the fractionalized fermi liquid. *Phys
Rev Lett, 105*(15), 151602. https://doi.org/10.1103/PhysRevLett.105.151602

Sachdev, S. (2010b). Strange metals and the AdS/CFT correspondence. *Journal of
Statistical Mechanics: Theory and Experiment, 2010b*(11), P11022. https://doi.
org/10.1088/1742-5468/2010/11/p11022

Sachdev, S. (2012a). What Can Gauge-Gravity Duality Teach Us About Condensed
Matter Physics? *Annual Review of Condensed Matter Physics, 3*(1), 9-33. https://
doi.org/10.1146/annurev-conmatphys-020911-125141

Sachdev, S. (2015). Bekenstein-Hawking Entropy and Strange Metals. *Phys. Rev. X,
5*(4). https://doi.org/10.1103/physrevx.5.041025

Teitelboim, C. (1983). Gravitation and hamiltonian structure in two spacetime
dimensions. *Physics Letters B, 126*(1-2), 41-45. https://doi.org/10.1016/0370-
2693(83)90012-6

White, C. D. (2021). Effective dissipation rate in a Liouvillean graph picture of
high-temperature quantum hydrodynamics. *arXiv.* Retrieved from https://arxiv.
org/abs/2108.00019v1

Witczak-Krempa, W., Sørensen, E. S., & Sachdev, S. (2014). The dynamics of quantum criticality revealed by quantum Monte Carlo and holography. *Nature Phys*, *10*(5), 361-366. https://doi.org/10.1038/nphys2913

Witten, E. (2016). An SYK-Like Model Without Disorder. *arxiv.org*, *hep-th*. https://arxiv.org/abs/1610.09758v2

Zaanen, J. (2019). Planckian dissipation, minimal viscosity and the transport in cuprate strange metals. *SciPost Physics*, *6*(5), 061. https://doi.org/10.21468/scipostphys.6.5.061